BATS

THEIR BIOLOGY AND BEHAVIOR

Tony Hutson

Comstock Publishing Associates
an imprint of
Cornell University Press
Ithaca, New York

First published in 2022 by the Natural History Museum, Cromwell Road, London
SW7 5BD
First published in the United States of America in 2022 by Cornell University Press

Librarians: A CIP catalog record for this book is available from the Library of
Congress.

ISBN 978-1-5017-6777-7

Designed by Mercer Design, London

Reproduction by Saxon Digital Services, UK
Printed by Toppan Leefung Printing Limited, China

Front cover: The painted bat, *Kerivoula picta* (see p.19). ©Merlintuttle.org/Science
Photo Library
Back cover: Flying fox, *Pteropus vampyrus*. ©Trustees of the Natural History
Museum, London.

Contents

Introduction

Bats are mammals, just like us. That is, they are warm-blooded animals with fur and they give birth to live young, which are suckled by their mother with milk. The bats form a single group of mammals, an order, called the Chiroptera (from the Greek 'cheir' meaning hand and 'pteron' meaning wing). The bats are one of about 30 orders of the class Mammalia, although with $c.1,400$ species they actually comprise about one-fifth of all mammals. The rodents include double that number of species and so account for about two-fifths of the mammalian species, with all the other orders including quite small numbers of species (the next largest is Primates, with fewer than 400 species).

The main feature of bats is that they are capable of sustained powered flight, which is made possible by elongation of the forelimbs, especially the hand (and hence the name Chiroptera). With the power of flight comes a range of other modifications as well as a range of opportunities not available to other mammals. There are also constraints. While their lifestyle is dominated by that power of flight, we will see that bats have evolved into a very successful and diverse group

OPPOSITE: Bate's slit-faced bat, *Nycteris arge*, from West Africa, one of 1,400 species of bat found around the world.

RIGHT: The brown long-eared bat, *Plecotus auritus*, has a large wing area that gives it highly manoeuvrable flight to seek out its insect prey.

of animals, demonstrating a wide distribution and a range of behaviour, diet and habitats. Like the majority of mammals, bats are generally nocturnal and in order to fly in the dark and find their way through obstacles and find prey they have developed a very sophisticated system of echolocation, the ability to produce and interpret the echoes from high-frequency (ultrasonic) sounds. This has also opened up new opportunities for resting sites, such as in deep caves.

Another very obvious feature of bats is that they spend much of their time hanging upside down. This involves some modification to their vascular system and other features. It also involves modification to the feet to ensure that the bats do not have to work to maintain themselves in their hanging position (in rather the same way that a bird's grip on its perch is automatically increased as it settles down to rest). The hanging habit enables bats to roost in relatively predator-free sites, but it also means that they do not have to push themselves up into flight: they can simply drop from their perch and take active flight once they have reached sufficient speed.

Horseshoe bats like this lesser horseshoe, *Rhinolophus hipposideros*, almost always hang free by their feet, other species may rest against the surface of their roost.

Unlike most rodents and other small mammals, which produce a lot of young over a relatively short lifespan, bats have the opposite strategy: they produce a small number of young (usually one per year, or not even every year) but live for a very long time (to a recorded maximum of 42 years). Thus, the baby bat is given a higher chance of survival at least until it becomes independent soon after weaning.

Some of the mechanisms for sperm storage in bats are also remarkable and unique in mammals. In temperate regions the males produce sperm during the summer and store it in the epididymis for mating in the autumn and, to some extent, through the winter. Once passed to the female, she will store the sperm in her reproductive tract, until she fertilizes the egg she produces when she ovulates following emergence from hibernation in the spring.

To some people bats may appear frightening and ugly, but their often strange facial appearance is linked to their particular lifestyle. It is really only in little more than the last 50 years that we have begun to understand much more about bats and their behaviour, much of it learnt through the availability of equipment that allows us to catch them, and to study them in the laboratory and increasingly in the field. Thus, these animals, which have been such a mystery to us and yet have often chosen to live in such proximity to us, are steadily giving up their secrets, secrets which sometimes we have found to be of great value to our own health and technical advancement. Our increased ability to spread knowledge about bats has removed a lot of the 'fear of the unknown' from people around the world, and has helped people understand the benefits that bats bring.

We can never know the full answer to the question that Thomas Nagel addresses in his classic philosophical essay What is it like to be a bat? and published in his *Mortal Questions* (1979), but we are very rapidly learning more and more about how bats work and what they do, and we can sample that knowledge in the following pages.

CHAPTER 1

Form and function

The basic structure of a bat is similar to that of other mammals, with the main obvious distinguishing features mostly being related to the ability of sustained powered flight and the sophisticated echolocation system associated with them flying and feeding in the dark. For much the same reasons, these animals are not renowned for their distinctive colouring and are generally quite dull, but there are some species that are spectacularly coloured or patterned, or adorned in other ways.

The wing

The wing is the single most obvious character of a bat, and yet it clearly retains the basic structure of a mammalian forelimb (see Chapter 2, Flight), and the flight membranes are layers of skin. Apart from the elongation of the forearm and fingers (digits 2–5), there is some modification of the elbow and wrist that

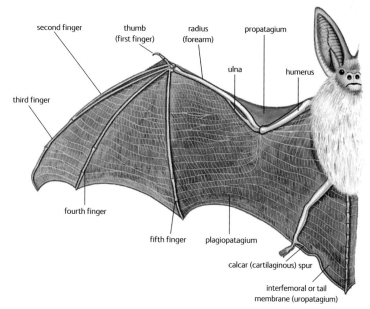

OPPOSITE: The northern ghost bat, *Diclidurus albus*, of Central and northern South America is one of the few white bats.

RIGHT: The structure of the flight membrane in bats.

second finger

thumb (first finger)

radius (forearm)

propatagium

ulna

humerus

third finger

fourth finger

fifth finger

plagiopatagium

calcar (cartilaginous) spur

interfemoral or tail membrane (uropatagium)

prevent twisting during flapping flight. In most bats the thumb is short but well developed and with a distinct claw used in climbing and grooming; in a few bats it is greatly reduced and almost clawless, and in others the thumb and wrist may have a distinct pad used for adhering to smooth plant surfaces (see Chapter 5, Roosts).

The head

The head, though, shows great variation. The general form of the head is based on the shape of the skull, which can be quite heavy depending on its musculature (and that is usually related to the kind of food, whether soft insects or fruit, or heavy beetles and other such prey). The muzzle of most bats is somewhat pointed, but particularly the nectar feeders of both Old World fruit bats and New World spear-nosed bats can be very elongate, taken to the extreme in a few of the spear-nosed bats (particularly *Platalina* and *Musonycteris*), in which the muzzle is several times as long as the brain case to give them access to the nectar of long tubular flowers.

At the other extreme are a small group of the fruit-feeding spear-nosed bats, which have an extremely broad flattened face; these also show wide secondary sexual dimorphism, both in colour and in that the females are very much larger than the males. *Ametrida*, the little white-shouldered bat, has some ancillary folds on the face, and the visored bat, *Sphaeronycteris toxophyllum*, has a large fleshy appendage protruding between the eyes, but these bats are completely outclassed by *Centurio senex*, the wrinkle-faced bat (or old-man bat), which, with its bare face with a complex arrangement of folds and deep furrows, must be the most bizarre-looking bat. That title is compounded by the bat having a fold of skin on the neck below the face that can be drawn up to cover the face like a visor; the skin is furred but there are two naked patches that can coincide with the position of the eyes. Further, there are translucent panels in the wings that enable the roosting bat to see any movement around it.

Just one other general feature of the head worth noting is that a number of bats have developed very flat skulls to enable them to squeeze into very narrow crevices, with extremes in the rock-dwelling African free-tailed flat-headed bats, *Sauromys* and *Platymops*, and the bamboo-dwelling *Tylonycteris* of Southeast Asia (see Chapter 5, Roosts).

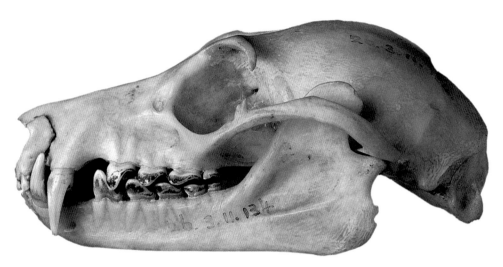

Skull of Old World fruit bat with extended muzzle and stronger development of cheek teeth (molars) for crushing fruit. Eye sockets are larger to accommodate large eyes.

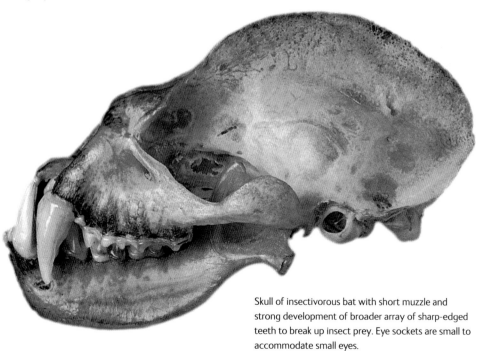

Skull of insectivorous bat with short muzzle and strong development of broader array of sharp-edged teeth to break up insect prey. Eye sockets are small to accommodate small eyes.

The teeth

Whilst it is not intended to go into the internal structures of bats here, the teeth are important. For the most part bats have retained most of the 44 teeth that are found in the basic (primitive) placental mammals, with 38 teeth in, say, the large genus *Myotis,* ranging down to about 30 in a wide range of species, but with the dentition drastically reduced to 20 teeth in the common vampire. The actual number of the four categories of teeth (incisor, canine, premolar and molar) varies greatly, as does their relative size and shape; only the canines remain constantly present as one prominent tooth in each upper and lower jaw, although they may vary a little in shape and size. Although there can be some individual variation in the number of teeth in a few bats (mainly Old World fruit bats), the left and right halves of the jaws are almost always the same; only in the lower jaw of the São Tomé collared fruit bat, *Myonycteris brachycephala,* the incisors are rather crowded and one is usually missing, leaving one prominent lower incisor and earning the species its local name of 'São Tomé one-toothed bat'. The teeth, then, are extremely important for species identification, and so too is the palate (the roof of the mouth) in the pteropodids, on which there is an array of ridges used in crushing fruit, and the number and arrangement of ridges is also important for identification.

The eyes

Most of the insectivorous bats – that is, most of the bats that rely on echolocation to find their way around and catch their food – have small eyes that are often partly hidden by fur. At the least, sight may be used to judge the light levels outside around the time of emergence, and it is likely that bats can discriminate some stars, and major landmarks or other features used in migration or homing; and some at least, such as the pallid bat (*Antrozous pallidus*), are able to discriminate patterns even at low light intensity. Indeed, it may be that bats' eyes are best adapted to low light levels. Other bats, such as some of the fruit and flower feeders, particularly of the New World spear-nosed bats, have more obvious eyes and can discriminate and recognize patterns. The eyes of the Old World fruit bats are all quite large and obvious and they are undoubtedly used for orientation and food finding. The ability to see in low light intensity is helped by the balance of rods and cones in the retina of the eye, but particularly in the fruit bats there are other mechanisms to help with visual acuity.

The ears

The variation in shape and form of the external ears of bats is tremendous. In most, the two ears are quite separate, while in a rather disparate range of bats, mostly big-eared bats, the ears are joined at the base by a fold of skin running across the top of the head; this is found in some molossids and vespertilionids, as well as nycterids, megadermatids and rhinopomatids (see Chapter 10, Classification for further details of these families). The most obvious part of the external ear is the pinna or conch, which can be short and rounded or extremely long, often with a row of transverse or sometimes longitudinal ridges, which probably provide support for the pinna but may have other functions. For the most part the ears are held erect, but in some molossids they are attached such that they lie almost pointing forwards. The ears of some species are almost as long as the body and when at rest some species can fold them down and tuck them under the wings. This keeps them out of the way, but when such bats are hibernating it can also prevent excess heat loss – while in one or two tropical species they are used to help lose excess heat. The ears of pteropodids are mostly rather unremarkable, but perhaps it is just worth noting that the base of the external ear forms a continuous ring where it joins the head, so that the ear is tubular. The ears are continually moving independently as the bats are monitoring their surroundings – even without echolocation. In the rest of the bats the ear does not form a complete ring but is open towards the front.

The ear has two other important features. Arising from just inside the base of the ear opening is a cartilaginous lobe that can be quite short and rounded or can be almost as long as the pinna. This is the 'tragus' and apart from the variation in length it also varies greatly in shape. The tragus is absent from pteropodids, rhinolophids and hipposiderids and often very reduced in molossids. The variation in the size and shape of the tragus has great value to systematists, but it must also have value to the bats, and it would seem that the tragus in conjunction with the shape of the pinna may be able to concentrate the sound receptive field to about 30–40° either side of the midline, and to improve the sensitivity to incoming echoes and the ability to discriminate their direction. While the determination of direction in the horizontal field is fairly easily judged by the time difference between echoes arriving at the ear, it is much more difficult to identify direction in the vertical field, and it is likely that the tragus helps here. In the rhinolophids, hipposiderids and molossids, where the tragus is absent or very reduced, there is a well-developed antitragus, which is a generally broad flap-like process that forms part of the outer margin of the base of the pinna.

The nose

The nostrils (or nares) are typically at the end of the muzzle and are rounded and directed somewhat to the side of the muzzle. On some species they are located on a 'narial pad', notably, as the name suggests, in Kitti's hog-nosed bat (*Craseonycteris thonglongyai*), and in the mouse-tailed bats (*Rhinopoma*). In the greater fishing bat and New Zealand short-tailed bat (*Mystacina*), the nostrils form a short slightly rolled tubercle, and this is taken further in the tube-nosed fruit bats (*Nyctimene* and *Paranyctimene*) and the tube-nosed bats of the genus *Murina*, which have, as the name suggests, the nostrils opening laterally at the end of short tubes.

In a range of bat families the nostrils are more or less incorporated into rather strange facial ornamentations, broadly termed noseleafs, and which are related to their style of echolocation. In the Old World false vampire bats (Megadermatidae) and New World spear-nosed bats (Phyllostomidae) a single blade-like (spearhead-like) noseleaf arises vertically from the equivalent of a narial pad around the nostrils. These noseleafs come in various designs, including becoming very reduced and distorted in the true vampire bats. The horseshoe bats have a much more complex noseleaf in which a horseshoe-shaped plate of skin more or less surrounds the nostrils (the 'horseshoe'). At the top this joins a triangular or subtriangular structure (the 'lancet') which has a more-or-less pointed tip standing erect behind the horseshoe and above the eyes, and usually with small pockets enclosed within the lancet. The third element of the noseleaf is a flat strap-like structure (the 'sella') arising from behind the nostrils and standing erect in the middle of the noseleaf. The sella is joined to the base of the lancet by a buttress-like 'connecting process'. Always a rather complex and curious structure, the noseleaf can be extremely ornate and bizarre, but would seem to be related to the echolocation system of these bats. The noseleaf of the Old World leaf-nosed bats is reminiscent of that of the horseshoe bats but more varied and can be even more complex. Some other families, such as the slit-faced bats, moustached bats and some of the free-tailed bats also have some strange facial structuring.

A range of strange bat faces. Clockwise from top left: common sword-nosed bat, *Lonchorhina aurita*; diadem leaf-nosed bat, *Hipposideros diadema*; wrinkle-faced bat, *Centurio senex*; eastern tube-nosed fruit bat, *Nyctimene robinsoni*.

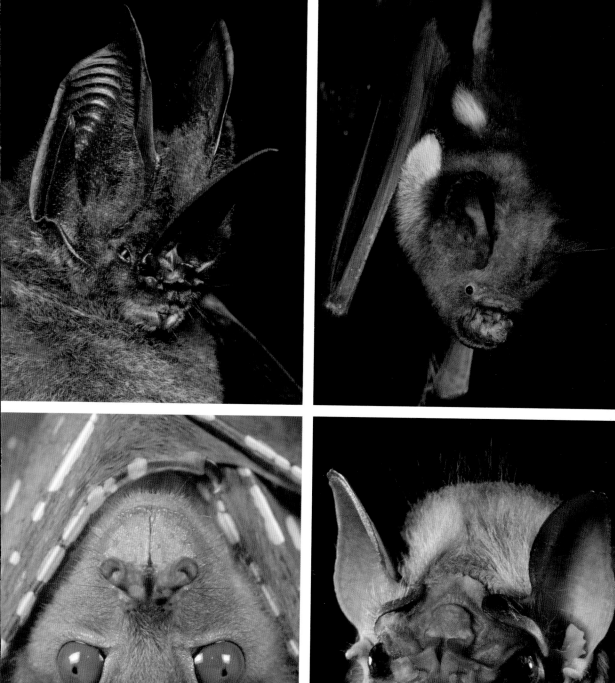
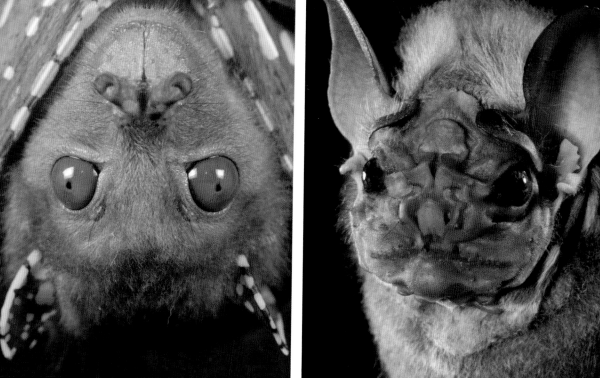

The body

The body varies little in shape, being more or less torpedo-shaped to benefit aerodynamics. It size does vary greatly, with the smallest bat arguably being Kitti's hog-nosed bat, or bumblebee bat, weighing between 2 g and just over 3 g (around 0.1 oz), but there are a few other species with weights that fall within that range. At the other end of the scale are the giant fruit bats of the genus *Pteropus* and *Acerodon*, with the largest reaching weights of 1.5 kg (3.3 lbs), such as *Pteropus vampyrus* and *Acerodon jubatus*. In other words, one individual of the largest fruit bats can weigh 750 times the weight of the smallest bat species.

External features of the body include paired thoracic nipples, usually one pair and present in both sexes. A few species have four nipples, and in one family the mammary glands are lower down the body towards the abdominal region. One interesting feature is a pair of false or pubic nipples present in the inguinal region just above the genitalia in the horseshoe bats and their relatives; these are not attached to the mammary glands and are mainly used as a holdfast for the young (see Chapter 4, Breeding). The male penis is external and obvious and is supported by a baculum (the penile bone or *os penis*, which is present in many mammal groups, including most primates, but not in humans). The testes are external and visible when a bat is undergoing spermatogenesis, but may not be apparent at other times. Even then, the caudae epididymides (ducts that store and transport sperm) are usually visible, lying either side of the base of the tail. Characters of the female nipples and male genitalia can give useful indications of the breeding status of many bats in the field, and the baculum (which is absent in all New World spear-nosed bats and some free-tailed bats and very reduced in some other bats) is also a very important taxonomic feature.

The hindquarters

The hindlimb and hindfoot have also undergone some major modifications associated with the flight of bats. The most fundamental shift is perhaps that the femur is rotated through 90° so that the knee points outward in comparison with other mammals (in the horseshoe bats and their allies the femur is rotated by almost 180°, but that is at the expense of their ability to crawl and climb). This rotation is related to the attachment of the flight membranes, both the wing membrane and the tail membrane (the interfemoral membrane or uropatagium) and the control of these membranes. This is discussed further in Chapter 2, Flight; however, it is perhaps worth looking here at the way bats hang upside down by their hindfeet. Apart from a few fish-eating bats, the feet of bats are generally quite small and with all five toes ending in a line. The weight of the suspended bat brings tendons into play that cause the toes and claws to hook onto any foothold. The way the toes and claws are designed means that the bat can sleep that way without expending energy. Even a dead bat will continue to hang in the same position. One other feature of the feet (or wrists) of some bats is the presence of pads to allow the bat to grip on the smooth vertical wall provided by a furled leaf of a banana or *Heliconia*, or the inside of a bamboo stem (see Chapter 4, Roosts). The sucker-footed bats of Madagascar (*Myzopoda*) and South America (*Thyroptera*) use an adhesive pad (moistened by secretions) to hold onto the substrate, whereas the pads on the feet of some other bats (such as the bamboo bats or banana bat) rely more on a microsculpture to provide adhesion – more like the foot of a gecko.

The relationship of the tail to the tail membrane varies greatly. As a basic pattern the tail membrane joins and is supported by the tail for all or nearly all of the tail length, i.e. the tail ends at or only slightly beyond the margin of the tail membrane. In the Old World fruit bats the tail membrane is absent and the tail is also absent or very reduced; only in the long-tailed fruit bat, *Notopteris*

A small vespertilionid bat showing the rotation of the femur so that the knees point upwards and to the side from the body.

macdonaldi, of some south Pacific islands is there a long free tail. The mouse-tailed bats (*Rhinopoma*) have a very narrow tail membrane, but an extremely long tail (almost as long as the head and body). The free-tailed bats similarly have much of the tail extended beyond the tail membrane, while in the sheath-tailed bats the tail protrudes from partway along the tail membrane when at rest, but in flight it is fully incorporated into the flight membrane. The New World spear-nosed bats show a mixture of long or short tail and short or long membrane, which is to some extent related to their diet – the tail and tail membrane being most reduced in some of the fruit feeding species and best developed in some of the purely insectivorous species.

Hair and colour

As hair is a unique feature of mammals, almost all bats have the body covered with a coat of hair or fur, which extends slightly and to a variable extent onto the wing or tail membrane. Only in the two species of naked bat (*Cheiromeles*, family Molossidae) is the body almost devoid of hair. In a few other apparently bare-backed bats of the Old World fruit bats (*Dobsonia*) and moustached bats (*Pteronotus*), the wings meet in the midline of the back but underneath, the body is furred as normal. The hairs are mainly fine short dense underfur or longer and coarse guard hairs. The hair itself consists of an inner medulla, and an outer cortex covered with many leaf-like cells, which are called cuticular or coronal scales. These scales on the guard hairs and the form of the hair roots can vary greatly and can sometimes be used for identification purposes. The length and texture of the fur also varies; for example, the fast-flying bats usually have short dense fur and some species, especially in the male, may develop long woolly ruffs or mantles. Specialized hair patches are found, particularly in males and associated with skin glands, such as on the shoulders of epauletted fruit bats, on the top of the head in Chapin's free-tailed bat (*Chaerephon chapini*), under the chin in some sheath-tailed bats, and often associated with the glands found on the wings of some bats. There is also a range of specialized sensory hairs found from the noseleaf of Old World leaf-nosed bats to the toes of free-tailed bats. Bats moult once per year, mostly in spring but with breeding females delaying moult until the young are weaned.

Although visual cues do not appear to play a strong role in the social behaviour of bats, there are still strikingly coloured or marked species. Most species are various shades of brown and grey. The tips of the fur are usually pale in colour with one or two (or more) bands of colour below the tip, and the pattern may vary with species or with age in some species. Some species, such as the hoary bat or parti-coloured bat, have an attractive frosted look to the back from white-tipped hairs scattered through the other fur. The American spotted bat (*Euderma maculatum*) and the African pied bat (*Glauconycteris superba*) have a very bold black-and-white pattern. And there are a few almost pure white bats, such as the sheath-tailed 'ghost' bat (*Diclidurus albus*) or the white bat (*Ectophylla alba*) of Central and South America. Amongst the most striking of bats are a few of the tropical *Myotis* species and picture-winged bats of the genus *Kerivoula*, in which the body fur and areas associated with the wing bones are bright orange, while the rest of the wing membrane is black. This pattern provides camouflage when they roost amongst dead leaves.

Another beautifully camouflaged bat is the little proboscis bat (*Rhynchonycteris naso*), which roosts out in the open on trees overhanging water; they are a striking mixture of greens and greys and with a lot of frosted tips to the fur giving them the appearance of a small bunch of lichen. This species and some of its relatives, the so called white-lined bats, have a pair of wavy pale lines down the back, which breaks up the image of a bat when it is resting on a tree trunk. Many of the Neotropical spear-nosed fruit bats have a line down the back and one or two lines on the face; the more prominent these lines are the more likely it is that the bat roosts mainly amongst foliage. Subtle patterns of fur colour occur in many other bats, many of which also roost in foliage where the fur colouring provides camouflage. With many of the canopy-roosting flying foxes, the patterns may relate to visual cues, which is very likely to be true for some of the more spectacularly patterned species, such as the Moluccan masked flying fox (*Pteropus personatus*) of New Britain and the spectacled flying fox (*Pteropus conspicillatus*) of New Guinea and northern Queensland, Australia.

The painted bat, *Kerivoula picta*, a vespertilionid bat from Southeast Asia, is brightly coloured, but well-camouflaged when roosting in dead leaves.

SOME INTERNAL FEATURES

Strong modifications are present in the skeleton, almost all of which are related to flight, with a basic requirement for reduced size and thickness of bones while retaining a strong structure. The rotation of the femur is discussed elsewhere, otherwise the main modifications are in the fusion of a variable number of the spinal bones to help maintain rigidity and modification to the thoracic skeleton for the change from the limbs being used for walking and running, to their use for flapping flight. These modifications include some fusion of bones, but also the development of attachment for a number of muscles used in flight. Indeed, the arrangement of muscles is quite different from that found in birds – which use very few, but very well-developed muscles attached to a deep sternum – whereas bats use about ten different muscles, with separate muscles powering the downstroke of the wing beat cycle from those powering the upstroke. There are also muscles that extend into the flight membrane and used for manoeuvring the membrane in flight.

Generalized skeleton of a bat.

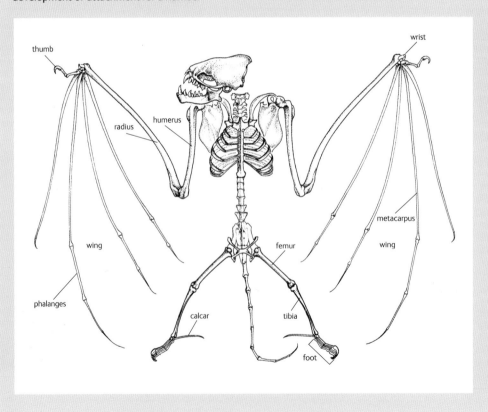

Again, many of the modifications of the circulatory system are related to flight, as well as to the use of daily torpor or for hibernation. The heart is very large and the heart rate of an active bat is significantly higher than in other mammals, but the heart rate can vary considerably depending on the activity of the bat and the ambient conditions. An active flying bat can reach a heart rate of up to 1,000 beats per minute, but during daily torpor that can drop to between 40 and 80 beats per minute, and during hibernation can drop to 10–20 beats per minute. The heart can pump an exceptional amount of blood during each stroke. Bats also have a unique ability to control the amount of blood in the wing membrane, such that they can limit the amount of blood circulating through the wing or through certain areas of the wing. This can be used for losing excess heat produced by flying or in high roost temperatures, or can be used to retain heat when a bat is resting or particularly in hibernation.

The lungs, too, have modifications to meet the high demand for oxygen during flight: they are relatively large and the alveoli are very small, leading to a very large surface area. Also the alveoli are highly vascularized with a high capillary volume in proportion to the alveolar surface area.

The digestive system of insectivorous bats is not so different from that of other mammals, but bats tend to have a short intestine that allows a rapid throughput of food (bats generally reject the hard and less useful bits of their prey before ingesting). In common with other mammals, the frugivorous bat species tend to have a longer gut but do have some special features. Basically, bats try to reduce the amount of weight they carry in flight, and this is best seen in vampire bats. The three species of vampire bats have modifications of the alimentary canal as well as the kidneys to enable them to take in a very large blood meal but then reduce the volume (i.e. remove the excess water) very quickly to enable flight. The kidneys of at least the common vampire then switch function back at the roost to limit water loss from body.

Another feature of bats is the enlarged larynx in all bats except the Old World fruit bats, i.e. in all the echolocating bats. The size and complexity of the larynx in bats is related to the volume and nature of the sound production. We shall see in Echolocation, p.29, that there are a few fruit bats that echolocate, but those that do, use a tongue-click mechanism that does not involve the larynx.

Perhaps more marked than the changes to the sound-production system is modification to the inner ear for sound reception and processing. Thus, there are various adaptions to the musculature and bone structure and perhaps most markedly with the cochlea and its associated labyrinths. This structure is larger and more complex in echolocating bats, it is not fused to the skull as is usual in mammals and is more helical than usual, indeed so much so that one scientist described a new species of snail from a cave in Crete, only to learn that what he had described was the cochlea of a greater horseshoe bat.

CHAPTER 2

Flight and echolocation

Bats are the only mammals capable of powered flight. Some other mammals, such as species of flying squirrel and the flying lemurs (or colugos), are capable of gliding long distances and of controlling their glide, but they cannot maintain controlled flight. Similarly, a number of reptiles and amphibians can glide, but are still only 'gliding' rather than 'flying'. So, from the vertebrates, it is only the bats and the birds that have mastery over the air and they, to a large extent, share it out with the bats being principally nocturnal and the birds principally diurnal.

Flight mechanism

Apart from the fact that both birds and bats flap their arms up and down to maintain themselves in flight, the flight mechanism is very different. In birds the 'arm' is relatively little modified – if anything, it is simplified – and the flight membrane is provided by rows of overlapping feathers. The bat retains the basic mammalian pentadactyl arm, but while the first digit remains a clawed thumb (and is useful for climbing and grooming), the other fingers, of digits 2–5, are extremely elongated with a fine elastic skin webbing stretched between the fingers (the dactylopatagium) and joining down the side of the body to the feet (the plagiopatagium). Usually the first and second finger (digits 2 and 3) are more or less joined at the end of digit 2, which gives some extra rigidity to the leading edge of the wing. In most bats the flight membrane continues from the hind legs to the tail (the interfemoral membrane, or uropatagium) and there is usually a narrow area of membrane in front of the forearm (the propatagium). The wing structure includes an elongated forearm (the humerus. then the radius with a very reduced

The European serotine bat, *Eptesicus serotinus*, a typical insectivorous bat, emits echolocation calls through its open mouth in flight.

ulna, which is not attached at its distal end) and the very long and fine digits 2–5. These digits support the wing membrane, which comprises two layers of skin either side of a network of blood vessels, thin muscles and a further network of elastic fibres that give the wings their flexibility in flight and assist with wing folding when at rest. To assist with the control of flight there is significant modification to the hind leg, such that the femur is rotated through 180° from the position that is normal in terrestrial mammals. Thus, the knees point what would be backwards in our case, upwards in the case of a bat. The leg bones are generally long and slender and not well designed to carry the weight of the bat, so that bats are usually good crawlers and climbers rather than walkers. In flight the legs seem to move up and down in a fairly passive way, but they help maintain balance and can be brought into more active play when the bat is carrying out manoeuvres as they are involved in changing the configuration of the plagiopatagium and uropatagium, and are involved in the bat's ability to execute stall turns. Most bats also have a cartilaginous spur from the heel (the calcar), which extends along the edge of the tail membrane for a variable distance towards the tail itself. This supports the trailing edge of the tail membrane, but is also movable to allow modification of the shape of the tail membrane in flight. The tail itself varies greatly in length and in the degree to which it is incorporated into the tail membrane. As we might expect, there is also significant modification to the musculature and skeleton of the thorax. In birds only two muscles – the pectoralis major and pectoralis minor – work antagonistically to power flight, whereas in bats many muscles are involved.

The net result is that bats have a large flight membrane; few bats soar or glide well or reach great speeds, but they are very agile and manoeuvrable. Bats with short broad wings have a low aspect ratio (the ratio of wing length to wing width) and are much more manoeuvrable, especially in slow or hovering flight, than those species with a high aspect ratio. The wing loading (the ratio of weight to wing area) is similarly important, with a higher wing loading requiring faster flight. The shape of the wing tip adds a further dimension. It might be thought that flight would be expensive on energy, but in fact bat flight is very efficient, especially taking into account the ability to cover great distances and to search for food. Another energy-saving practice is that bats tend to drop out of their roost site and then, once they have gained enough speed, use flapping flight to get the lift required to stay airborne. Some of the long and narrow-winged bat species (those with a high aspect ratio) may drop as much as 5–6 m (16–20 ft) before opening their wings. That's different from the burst of energy required by birds to take off from the ground or a perch.

PTEROSAUR

A comparison of the wing of a pterosaur, a bird and a bat.

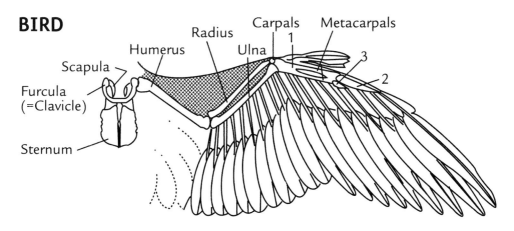

Pteroid
Radius
Humerus
Ulna
Carpals
Metacarpal
1 2 3 4
Scapula
Coracoid
Sternum

BIRD

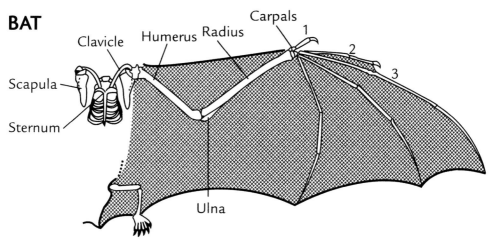

Carpals
Radius
Humerus
Ulna
Metacarpals
1 2 3
Scapula
Furcula (=Clavicle)
Sternum

BAT

Carpals
Clavicle
Humerus
Radius
1 2 3
Scapula
Sternum
Ulna

Efficiency

Flight does cost energy, but comparison with birds and hawk moths suggests that bats are relatively more efficient. There are many specific physiological problems associated with flight, such as lung ventilation and thermoregulation. Bats can breathe in time with the wingbeats, which saves energy, and another energy-saving mechanism is to use that breathing to manage the emission of their echolocation calls, so that the action of the wings means that the breathing and emission of echolocation calls are free of extra energy costs in normal flight. Flight also creates excess heat, which could be a problem, but it is worth noting that, for example, the flight membranes of a pallid bat make up approximately 80 per cent of its surface area, and so heat loss can be managed through the flight membranes. However, when a bat is hibernating it does not want to lose too much heat (or fluid) through evaporation through the flight membranes, so the membranes are usually tightly folded – and even the large ears of *Plecotus* and *Corynorhinus* species are folded out of the way.

There are many other physical or biomechanical issues associated with flight in bats to improve their efficiency. Some of these may seem very minor but are actually quite important. Thus if you sit overlooking the wing of an aeroplane, you will probably see one or more rows of studs running along the wing. These create microturbulence in the air flowing over the wing surface, which helps keep the boundary layer of air on the upper surface of the wing and contributes to providing the lift given by the air flowing under the wing surface. Bats have hairs along the leading edge of the wing, but more particularly the knobbly knuckles of the finger joints appear prominently above the upper surface of the wing membrane; these features appear to do the same job as the studs on a plane's wing (or, we should say, vice versa).

For the most part bats are not racers. Most small insectivorous bats generate airspeeds of between 18 and 30 km/h (11 and 18 mph). A gleaning bat hunting in clutter may be slower than that, while open area foragers with longer narrow wings, such as the noctule bat or European free-tailed bat, *Tadarida teniotis*, are faster at up to 50 km/h (31 mph). If one of these bats rolls into a vertical dive after an item of prey, it undoubtedly travels faster. Indeed, other free-tailed bats are even faster, with the Australian white-striped free-tailed bat, *Austronomus australis*, topping 60 km/h (37 mph), and Brazilian free-tailed bat, *Tadarida brasiliensis*, has been recorded with a maximum speed of more than 160 km/h (99 mph) in level

LEFT: The short broad wings of the brown long-eared bat, *Plecotus auritus*, are good for slow, hovering flight within a cluttered environment.

BELOW: The long narrow wings of a bent-winged bat, *Miniopterus* sp., from Angola are good for fast flight in open territory with few obstacles.

The common vampire bat, *Desmodus rotundus*, is capable of leaping off the ground in order to take flight – a sort of vertical take-off.

flight. While the mass emergence of some of the large colonies of free-tailed bats is spectacular, perhaps more so is their return at dawn, when at many caves the bats accumulate hundreds of feet above the cave mouth and then, one by one, close their wings and plummet in free fall into the cave mouth, where they immediately open their wings to negotiate their way within the cave. The maximum speed during that free-fall drop into the cave is likely to be faster than they are capable of in level flight, and the deceleration at the end of the fall is also an amazing feat.

Bats are generally less agile on the ground than most other mammals, and while generally capable of taking off from level ground, a lot of bats have difficulty with it. The most agile bat on the ground is undoubtedly the common vampire bat, which can move very rapidly in any direction and can readily spring off the ground into flight. The New Zealand short-tailed bat, *Mystacina tuberculata*, spends quite a lot of its time moving through leaf litter on the ground searching for invertebrate prey, while it also searches for the ground-flowering wood rose, *Dactylanthus taylorii*, for which it is a prime pollinator. *Murina* species also seem to spend a lot of time on the ground, and many other bats, especially free-tailed bats, are pretty good 'walkers'. The horseshoe and Old World leaf-nosed bats are less adapted to walking on all four limbs than most other bats, although they are very capable of 'walking' (using just the hindlimbs) across the roof of a cave or other roost. Most bats are also quite good swimmers, although not for long distances. In addition, they cannot easily take off from the water surface and so need to get to shore fairly quickly.

Echolocation

How bats find their way around in the dark had long been a mystery. In 1793, an Italian scientist called Lazarro Spallanzani experimented with bats and found that blinded bats could fly well and catch food, but deafened bats could not. At about the same time, Charles Jurine, a zoologist in Switzerland carried out some rather less traumatic experiments to demonstrate the same thing – indeed Spallanzani and Jurine collaborated at some level. But their research didn't answer the question 'How?' and so was widely ignored. It wasn't until 1932 that a Dutch zoologist, Sven Dijkgraaf, recognized an auditory basis for the detection of obstacles by bats (he could hear the calls of bats reacting to obstacles), but had difficulty demonstrating it to the wider world. At this time George Pierce, in the Physics Laboratories of Harvard University, was developing electronic apparatus to study the sounds of insects, including to the ultrasonic level. In 1938, Donald Griffin of the Biology Department of the same university, took him a couple of bats. They flew the bats in a room and it was immediately obvious from sounds recorded by Pierce's apparatus that the bats were producing a lot of ultrasonic sound and using this to navigate. In an appendix to his book *The Songs of Insects*, Pierce says that in that year (1938) 'Dr Griffin and I published a brief note on this discovery'. That 'brief note' was the start of a whole new field of investigation and equipment development. With Robert Galambos, a physiologist of the Harvard Medical School, Griffin continued laboratory investigations into the echolocation process in bats and published a number of key works, including his classic *Listening in the Dark*. Soon there was a call to develop mobile apparatus that could be taken into the field. Thus the first field 'bat detectors' that could convert the ultrasonic sounds of the bat to something that could be heard in the field were created in the early 1960s in the UK.

Basically, most bats in flight are emitting a stream of high-frequency sounds from the larynx, mostly of a higher frequency than we can hear (hence 'ultrasonic') and building up a 'sound picture' of their surroundings from the echoes received.

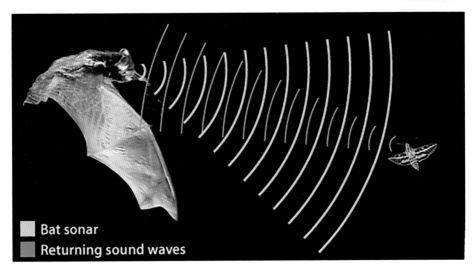

Bat sonar
Returning sound waves

A bat beams out ultrasonic signals which, on encountering an object, return
echoes to the bat's ears.

The amplitude, frequency, pulse repetition rate and call structure can all vary with species and behaviour. This is a fascinating but hugely complex subject and beyond the scope of this book to go into any detail. However, given here is a general overview of the distribution of echolocation in bats, the sound production and its reception and how it is used for navigation, foraging and feeding. The response of prey is discussed elsewhere (see p.45).

DISCOVERY

It was assumed that very quickly bat detectors would be able to identify all the bat species flying by and explain what they are doing, but life is not so simple. Despite continuing improvements in technology, in the receiving and analysis of recorded sounds, and despite the claims made of many machines to discriminate all species of bats in the field, accurate automatic identification in the field is still far from being achieved. Even in Europe the accuracy of automatic identification can be as high as 80 per cent or more or as low as 30 per cent, and so the identifications still need to be checked. The problem is that the echolocation calls of bats do not serve the same function as the songs of insects or birds – they are not primarily for identification and advertisement. They are there to assist the animal in finding its way around in the dark and to locate its food. Thus although there are fundamental differences between the calls used by different groups of bats, the calls change depending on such factors as whether a bat is commuting or searching for prey, feeding (and the stage in the location and capture of prey), the nature of the prey, and the habitat that the bat is flying in.

Nevertheless, the development of bat detectors and other equipment has resulted in many major advances in the study of bats, and as long as we understand the limitations of the detectors, they have greatly helped with field identification (and will get better). Underlying that are huge advances in the understanding of echolocation, how it works and the links between the diversity of bats and the way they use echolocation (suggesting in some cases that echolocation may have evolved more than once in bats). Until the 1950s, most information on bats was gained from bats in their roosts; now, our ability to study, survey and monitor bats in the field gives a very much more complete picture of their distribution, behaviour, activity and habitat use, and hence a much better opportunity to contribute to their conservation and assess the impacts of land management on them.

WHICH BATS PRODUCE WHAT

The Old World fruit bats (Pteropodidae) do not produce ultrasonic laryngeal echolocation calls, but a few species have developed other mechanisms to find their way in and out of cave roost sites. Some cave-dwelling fruit bats, such as the lesser dawn fruit bat, use fairly low-frequency wing clicks (wing clapping),

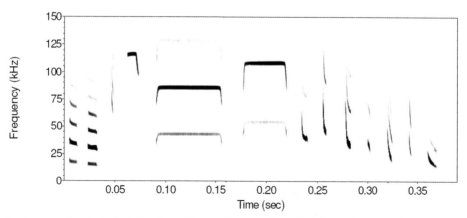

Spectrogram of one typical echolocation call from each species of a guild of thirteen bat species foraging in acacia woodland in Israel. From left to right: *Rhinopoma hardwickii, R. microphyllum, Nycteris thebaica, Asellia tridens, Rhinolophus hipposideros, R. clivosus, Pipistrellus kuhlii, Hypsugo bodenheimeri, Eptesicus bottae, Barbastella leucomelas, Otonycteris hemprichii, Plecotus christii* and *Tadarida teniotis.*

audible to some human ears, as a rather simple means of finding their way around caves and locating walls to roost on, but not giving detailed enough information to detect small obstacles or precise distances. Some other fruit bats, such as species of *Cynopterus* and *Macroglossus* use a primitive low-frequency tongue-clicking mechanism, perhaps similar to the system used by oil birds and cave swiftlets, to achieve the same ability. Only within the genus *Rousettus* several species use a much more sophisticated tongue-clicking mechanism ('lingual echolocation'), but at a much higher frequency (into the ultrasonic at up to 30 kHz), to find their way around cave roosts, where they can avoid small obstacles and hook onto small perches with precision.

All other families of bats use laryngeal echolocation, i.e. the calls originate in the larynx and are beamed ahead of the bat. Most species shout through the mouth, but some, such as horseshoe bats, plecotine bats and New World spear-nosed bat, shout through the nose. Sometimes the sound is quite directional, as in the horseshoe bats; in most cases it is a much broader cone. The lower the frequency the further the sound will travel, but the less information is acquired. The African free-tailed bat, *Otomops martiensseni*, echolocates at *c.*10 kHz and mostly flies high and fast in the open, feeding on moths, whereas Percival's trident bat, *Cloeotis percivali*, also from Africa, is generally credited with using the highest frequency at over 200 kHz; this species also feeds on moths, but usually within the cluttered vegetation of woodland. The fisherman bat is often credited with being the loudest bat in the world, capable of producing sounds at over 140 decibels, while the 'whispering bats' of the genera *Plecotus* or *Corynorhinus* may be so quiet it is difficult to detect them from any distance away.

Components of the calls of bats are usually separated into constant frequency (CF) and frequency modulated (FM). As the names suggest, the CF calls are more-or-less long calls on a particular frequency, whereas FM calls sweep very rapidly down through a broad range of frequencies. CF calls give better information on the texture of the surroundings (including the presence of an insect resting on vegetation) and allows accurate range estimation of a target. The CF calls allow the use of the Doppler effect (Doppler shift) to estimate target range and to give information on the wingbeat frequency and other details of flying insects. The Doppler shift is the change in pitch (frequency) of a moving sound source relative to the position of the receiver. Thus the pitch of an echo rises in an approaching insect and lowers in an insect moving away. We are familiar with Doppler shift from, say, the sound of a passing train, with the rising pitch of the sound as the train approaches, and falling pitch once it has passed. Other variations of the calls include the frequency range, their repetition rate, rhythm, the loudest frequency and the volume. Horseshoe bats and their relatives mostly use CF calls, but usually with a small FM 'tail' at one or both ends. *Myotis* species and plecotine bats almost exclusively use FM calls, and many other species use a combination of the two, usually the call starting with an FM component and

Spectrograms of a selection of calls of noctule bat, *Nyctalus noctula*: 1-4 echolocation calls, 5-6 social calls.
1. foraging high above riparian habitat; 2. foraging above street lights; 3. foraging over riparian treeline;
4. flying very low under canopy of tree; 5. social call of stationary male; 6. social call of male in flight.
(Time (in ms) is shown horizontally, frequency (in kHz) vertically, intensity of sound (volume) increasing from red to green.)

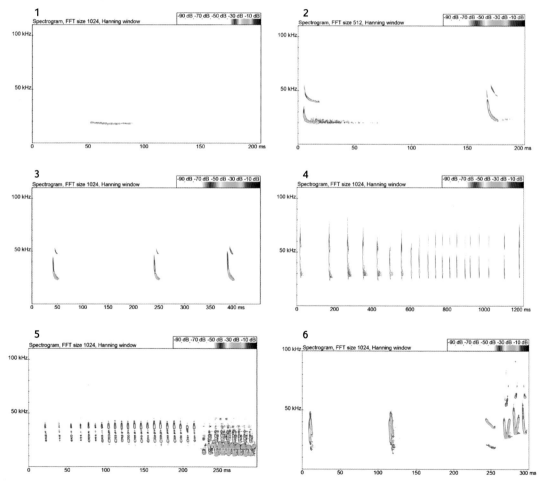

finishing with a short CF component. In addition, there is the variation that an individual bat will use in its various activities. Thus, a noctule bat waiting to emerge from its roost will give an audible chatter and sometimes loud shrieks. Once emerged and in its hunting mode, it alternates a shallow FM sweep with a short CF call – as it flies along it makes a call that sounds like 'chip-chop – chip-chop – chip-chop', but as it first drops out of the roost it will only use the FM part, and if it is flying high in the open commuting to a feeding area it will only use the CF part at a slow repetition rate. Closer to the ground or when feeding around street lights it drops the CF part altogether and increases the pulse repetition rate for the FM part, particularly as it homes in on insect prey. As it enters a woodland to return to, say, a tree roost, it relies more and more on the FM part, and when it is finally swarming around outside the roost site and waiting to enter, the fast repetition rate and almost pure short FM call can be confused with quite different bats, even *Myotis* species. In open commuting flight a bat might emit sound pulses at less than ten pulses per second, but in the final stages of approaching a prey (the 'terminal' phase), the length of the call decreases and the pulse repetition rate increases to as much as 200 pulses per second.

The nature of the calls has a considerable relationship to the range of facial features visible in bats. Thus the facial adornments, particularly of horseshoe bats and their allies and New World spear-nosed bats, are related to the echolocation signals they use. These structures appear to be related to the frequency and directionality of the calls, and probably serve other functions. Similarly, to a considerable extent, variation in the external ears is all to do with the reception of the information returning to the bat for its subsequent interpretation in the inner ear. Bats have a range of sophisticated techniques to prevent deafening themselves when emitting calls, to interpret the (often very faint) echoes that return, and to isolate the information that relates to their own activity from that of other bats. Bats also produce a range of other calls, generally termed 'social' calls. These may relate to territoriality or advertisement for mating, but an increasing number of calls are being identified as functioning for communicating information, particularly within a maternity colony between mother and young.

While the use of sound is such an important feature of the majority of bats, other senses such as sight and smell are also important. Sight is much more important to bats than the general public perceive, especially to the fruit- and flower-feeding Old World fruit bats and the New World spear-nosed bats, for surveillance for predators, general navigation and locating feeding areas. Additionally, a range of insectivorous

The California leaf-nosed bat, *Macrotus californicus*, finds much of its insect prey by sight.

bats capture a part of their diet by sight, especially those gleaning bats with large ears and broad wings that take resting prey; indeed, the big-eared leaf-nosed bats of the genus *Macrotus* may take most of their food by sight. The use of smell is most developed in the fruit and flower feeders for the location of their food, for its identification and to assess its state of ripeness. Of course, the plants themselves help in this respect in dispensing aromas to attract the bats. Smell is also associated with scent glands that are quite widely distributed in bats and may provide a variety of functions, including territory marking and advertising mating potential.

CHAPTER 3

Food and feeding

About 75 per cent of all bat species are insectivorous. Nevertheless, bats exhibit a wide range of food and foraging behaviour. In the tropical and subtropical Old World, from West Africa through to islands of the western Pacific, one family of bats, the Pteropodidae or fruit bats, comprising around 170 species, feed on fruit and flowers and sometimes on significant quantities of foliage. Similarly, in Central and South America a large part of the family of spear-nosed bats, again comprising around 170 species, are mainly fruit and flower feeders. However, this family also includes a small number of bat species that are largely carnivorous, including some that will take other bats, or are fish and crustacean feeders; and then there is the small group of three vampire bats of Central and South America that feed on blood. Although there are no 'grazing' or confirmed 'carrion' bats, the range of diet exhibited by bats fairly well parallels the other terrestrial mammals and the birds.

OPPOSITE: The hairy big-eared bat, *Micronycteris hirsuta*, of Central and South America feeds mostly on large insects.

LEFT: Marianas flying fox, *Pteropus mariannus*, feeding on fruit.

Insectivorous

For the insectivorous bat species, as we have seen, the sophisticated echolocation that varies from species to species and can be varied by the individual bat, gives bats mastery of the night sky and access to the wealth of insects and other invertebrates that are active at night. Some bats dart around after small midges and other such insects and may take 2,000–3,000 insects per night. Other bats lead a more relaxed life, using a 'feeding perch' to wait for a passing large beetle and then, like a flycatcher, darting out and returning to their perch with the prey. For convenience, the foraging behaviour of bats is divided into about five different strategies. Although most species have a favoured foraging strategy, many can utilize more than one technique. Probably the widest-used technique is 'aerial hawking', the taking of prey in flight, often in relatively open habitat, where the detection of prey is easier with less confusing background echoes. Some species, such as *Pipistrellus* and *Myotis*, tend to stay close to the edge of vegetated areas. Others, particularly larger species and species with longer narrower wings, such as *Nyctalus* and *Lasiurus* species, as well as free-tailed and many sheath-tailed bats, can fly fast in the open and often very high in the sky. We can include as aerial hawking, the practice of bats such as the noctule bat (*Nyctalus noctula*) suddenly to spiral down towards the ground to take an insect it has detected below it (sometimes called 'stoop feeding'). Brazilian free-tailed bats (*Tadarida brasiliensis*) exploit high-flying swarms of migratory insects such as corn earworm moths (*Helicoverpa zea*). While the species of bat and insects remains as yet undetermined, bats have been recorded by radar as probably feeding amongst migratory swarms of insects as high as 1,220 m (4,000 ft), and possibly up to 1,830 m (6,000 ft) above ground level.

Within a cluttered habitat, such as the crowded vegetation of woodland, many bats, including *Myotis* species and horseshoe bats, still use aerial hawking as a means of detecting and catching prey, but they modify their echolocation signals to ensure that they do not receive too many confusing echoes. Many species foraging in clutter will also prey upon invertebrates resting on a leaf or a tree trunk; indeed, a number of species specialize in this 'gleaning' technique. Apart from gleaning insects from vegetation, some species glean spiders from their webs. This is particularly a feature of some *Myotis* species.

Perhaps related to gleaning is 'pounce feeding', whereby the bat flies very close to the ground and may detect prey partly by echolocation, but also by listening

The New Zealand short-tailed bat, *Mystacina tuberculata*, here approaching flowers of the endemic woodrose, which it pollinates. It also seeks insects in the leaf litter.

A pallid bat, *Antrozous pallidus*, is one of several species that can prey on scorpions.

for the movements of the prey itself. The bat will either drop straight onto the prey, or land very close to it and chase it over the ground. Thus a large part of the prey of species such as the European larger mouse-eared bats is flightless ground beetles, bush crickets and other non-flying arthropods, such as centipedes. The smaller European Bechstein's bat (*Myotis bechsteinii*) and the big-eared bats of the genera *Plecotus* (Europe) and *Corynorhinus* (North America) all have very large ears and are also great passive listeners, using the sounds of their prey to enable them to sneak up on them and use gleaning or pouncing techniques. In more arid regions, other large-eared bats such as *Antrozous* in North America and *Otonycteris* in North Africa use much the same techniques to catch their prey, which will often include invertebrates such as scorpions. Even more terrestrial are the tube-nosed bats of the genus *Murina* and probably the New Zealand endemic *Mystacina*, which spend a significant amount of their foraging time scrambling around in leaf litter and the like, searching for invertebrate prey.

The technique of 'trawling' or gaffing prey from the surface of water is practised particularly by a range of *Myotis* species from around the world. This involves flying low over the water to detect prey such as insects emerging on the water surface. These bats are characterized by having very big feet with large claws, which may be distinctly bristled. Having detected a suitable prey item, the bat will drop its feet and/or tail membrane into the water to drag the prey from the water surface and will then pass it to the mouth in flight. The prey is usually eaten in flight before the next feeding attempt; meanwhile the bat continues to produce echolocation calls. Depending on the species involved, they generally prefer the still flat stretches of rivers and lakes to feed in, but they can also be seen doing the same thing over the sea, where they may be taking insects that have been blown out to sea, but may also take small crustaceans that accumulate at the water surface. A group of researchers from the Basque Country led by Joxerra Aihartza has shown that the European long-fingered bat, *Myotis capaccinii*, can use the same technique to catch small fish, but fishing is taken to a more professional level in bats from Central and South America (see p.42).

The use of 'perch-feeding' or 'flycatching' is found in a range of bats, especially many horseshoe bats and leaf-nosed bats, but also a number of other broad-winged bats, such as the Old World false vampires (Megadermatidae) including *Megaderma* and *Lavia*, and the slit-faced bats (Nycteridae). Hanging from a suitable vantage point (usually from vegetation) these bats monitor their surroundings for the passing of a suitable prey item. Having located prey, they drop from the perch and chase it. Once caught they take the prey back to the original perch to eat it or take it to some other regularly used feeding perch, where the prey will be eaten and where wings and other hard parts may be discarded. A regular feeding perch is also a feature of many gleaning bats. The discarded remains ('rejectamenta' as it was known) can be a very useful source of information on the diet of such bats.

Eastern Daubenton's bat, *Myotis petax*, trawling for pelagic crustaceans that sometimes swarm at the water surface of Lake Baikal, Siberia.

Carnivorous

For the most part, the small numbers of carnivorous (or more usually partially carnivorous) bats are those with slow manoeuvrable flight that use the gleaning/perch/pounce type of foraging strategy. An exception is the European greater noctule, *Nyctalus lasiopterus*; this is a bat with long, relatively narrow wings that usually forages on insects in open spaces, but during bird migration time this bat feeds to a great extent on small night-flying migratory passerine birds, such as willow warblers and robins. Quite how these bats catch and eat their prey is not well known (a bat would not want to fly too far carrying a robin that weighs about half its own weight). Also, for digestive reasons, these bats do not find it easy to eat pure meat, and while they may pull off the wings, the faecal pellets contain a disproportionate amount of feathers, which must provide some roughage to the bat. During the peak migration period as many as 70 per cent of the droppings include quantities of feathers.

As with the greater noctule, the carnivorous bats are perhaps more correctly only partial carnivores – they all take some insect food at times. In reality, it is perhaps a small step from taking a large insect to taking a small lizard or frog. Hence most of the groups discussed above under the gleaning or perch-feeding category include species that are to some extent carnivorous and use passive acoustic cues, such as the chirping of frogs or movement through vegetation, to locate their prey. In the New World there are carnivorous members of the spear-nosed family, including the greater spear-nosed bat, which will take a range of small vertebrates including other bats. In addition to insectivory and carnivory, this bat also eats fruits and so is actually a true omnivore. As its name suggests, the frog-eating bat, *Trachops cirrhosus*, is more of a specialist and homes in on the singing of frogs to locate its prey.

As its name suggests the frog-eating bat, *Trachops cirrhosus*, is a specialist feeder on small frogs and toads.

Fishing

Also in the New World we find the most specialist fishing bats. One of these is *Myotis vivesi*, the fish-eating myotis. This bat belongs to a huge genus that is found throughout most of the world and shows considerable variation in form and behaviour, but this species is perhaps the most extreme and so is often included in its own subgenus *Pizonyx*. It is a rather large species of *Myotis* with huge hindfeet and very long toes and claws. It is restricted to some coastal parts of the Gulf of California and adjacent parts of the Pacific coast of Mexico. It roosts amongst boulders, scree, in caves and rock crevices and forages over the sea for fish and crustaceans, but it will sometimes take insects and even algae (although it is not entirely clear that the algae is taken intentionally).

Another fish specialist is the greater bulldog bat, or greater fisherman bat, *Noctilio leporinus*. This is one of the largest bat species of Central and tropical South America (including parts of the Caribbean) and is characterized by its drooping jowls and its very long legs with enormous feet and claws. It trawls

The greater fisherman bat, *Noctilio leporinus*, trawls for fish in the still waters of lakes, streams and even the sea.

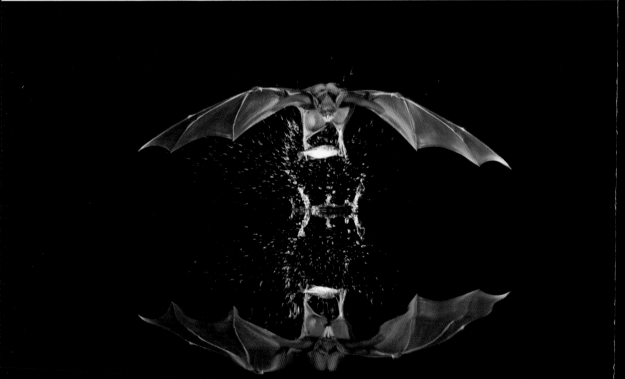

for fish from the surface of calm waters of harbours, estuaries, lakes and rivers and will often specialize on particular fish species. The fish are caught either by targeted dipping of the feet into the water, or by just raking the water surface for c.10 m (33 ft). Fish predominate in this bat's diet, but it will also eat crabs and shrimps and a wide range of other invertebrates, including soft-bodied insects such as flies. This and one other species make up the small but very distinct family Noctilionidae. The lesser bulldog bat, *N. albiventris*, is similar in appearance but a little smaller than its relative and very much less of a fish specialist, rather eating a wide range of invertebrates as well as some fruit; it is absent from the Caribbean.

The pallid bat, *Antrozous pallidus*, perhaps demonstrates the most varied diet. This quite large vespertilionid mainly feeds on invertebrates, especially grasshoppers and allies, beetles, sun spiders, spiders, scorpions and occasionally vertebrates such as small lizards, rodents and other bats. But its diet shifts both seasonally and geographically – in some areas it becomes an important pollinator of some cactus species and is the only nectar-feeding vespertilionid. It will also feed on other flowers and fruit, but perhaps with more interest in the insects within.

Blood-feeding

Before looking at the fruit and flower feeders in more detail, we should look at the extraordinary sanguinivorous (blood-feeding) vampire bats. A legend of vampirism existed in eastern Europe, based upon the theory that some dead persons could leave the grave to drink the blood of the living, long before the blood-feeding bats of Central and South America were discovered. The association of bats and vampirism eventually came full circle when Bram Stoker published his eternally popular novel *Dracula*. But true vampire bats rarely feed on humans, and it was with some disappointment that eccentric English naturalist and explorer Charles Waterton, travelling around Guyana in the early 1800s with a Scottish colleague, reports in his *Wanderings in South America* (1825) that he never got 'visited' by a vampire bat (despite his best efforts to 'tempt this winged surgeon') although his colleague did.

There are three species of vampire bat, all restricted to Central and South America. They do not 'suck' the blood from their host, but rather graze the skin with their highly modified incisors and canines to allow the blood to flow, and then they lap it up, or the blood flows along grooves in the tongue into the mouth, and their saliva includes an anticoagulant (called Draculin) to maintain

Common vampire bat, *Desmodus rotundus*, feeding
from a chicken's toe.

the flow of blood at least until the bat has finished feeding. One species, the
common vampire (*Desmodus rotundus*) feeds mainly on mammals, the hairy-
legged vampire (*Diphylla ecaudata*) feeds mainly on birds and the white-winged
vampire (*Diaemus youngi*) is thought to have a preference for birds.

Vampire bats are not large bats, with an 'empty' weight of *c*.25–45 g (0.9–
1.6 oz) and a forearm of 50–60 mm (2–2½ in), but they can imbibe a relatively
large quantity of blood. The common vampire, which spends more time on the
ground than the other two, will sometimes sit quietly waiting for its kidneys to
remove water from the blood meal before it loses enough weight to be able to
fly. A unique feature is that, at least in the common and hairy-legged vampires,
the bats may return to their colony with excess food and can be encouraged to
feed an unfed youngster or older animal of its species. Most of the observation
of feeding of these bats is on domestic animals, while the natural food has been
little recorded. But it is likely that the common vampire has become much more
common with the introduction of suitable food sources such as cattle, horses and
pigs. Nevertheless, its natural food sources are likely to be quite varied, as they

include such species as tapirs in forest and sea lions in coastal or island situations. The common vampire bat is also extremely fast and agile on the ground, which enables it to approach its host unnoticed and to take emergency action to back off if the host moves. Neither of the other vampire bats spends much time on the ground, but the white-winged vampire sometime uses a cunning trick to help it feed on chickens – if the bird seems to react to the bat's attempt to feed, it will nuzzle under the chicken like a chick and wait until the chicken has settled on its would-be youngster before the bat starts to feed.

For the insect-eating bats there are problems in finding and catching their prey. Many insects have 'ears' and so can hear a standard aerial hawking bat as it approaches. These insects will react to avoid being taken, usually by dropping out of the sky or by wildly erratic flight. Many of the gleaners get round that by using very quiet echolocation calls (they are the so-called 'whispering bats') or passive listening to enable them to sneak up on their prey. Similarly, the 'pounce-feeders' use little or no echolocation calls so that they too can catch prey that would otherwise be warned of their approach. There are also insects that answer back when they hear a bat coming. Some arctiid moths produce a loud 'harsh' sound when they hear a bat approaching; this may warn the bat that the potential prey is distasteful, or may just confuse the bat for long enough for the insect to take evasive action. Some bats use echolocation calls that are outside the reception frequency of the tympanate moths (the tympanum being the hearing organ of these moths) that they feed upon, thus free-tailed bats such as *Otomops martienssenii* and *Tadarida teniotis* (and the long-eared vespertilionid *Euderma maculatum*) use frequencies below that of the moths' hearing, while horseshoe bats and Old World leaf-nosed bats use frequencies above the moths' hearing.

So, generally, insectivorous bats increase their access to moths that can hear echolocation calls by using frequencies to which the moths are less sensitive. But in some areas, such as Africa, the frequency range of hearing in the moths has become extended to enable the avoidance of a wider range of bat species. Thus the prey defences may be as important in determining the structure of a bat community as competition between bat species. On the other hand, in woodland, the eared moths may have a shorter predator-detection distance and coupled with a cluttered habitat may not have enough warning of an approaching bat. Changing the pattern and frequency of echolocation calls during the pursuit of prey and the use of quiet or 'stealth' calls are further adaptations by the predator to increase their catch of eared prey.

FRUIT- AND FLOWER-FEEDING

The fruit- and flower-feeding bats occur in two families: the whole of the family of Old World fruit bats, Pteropodidae, and a large part of the New World family of spear-nosed bats, Phyllostomidae. Both groups include about 170 species. As we have seen, there are some other species that will feed on fruit or flowers at some time, but these two groups are the specialists. Despite being quite unrelated, they do demonstrate remarkably similar morphology and feeding strategies. In the Pteropodidae, the tail membrane is absent or reduced, the eyes are well developed and the nectar feeders tend to be small species with an elongate muzzle and a very long tongue, often with papillae at the end to help catch the nectar. More specific modifications for nectivory may be related to the particular choice of flowers. The family lacks echolocation except for a rather primitive tongue-clicking technique used by one small group of species (the Rousettinae) to find their way around caves. The fruit and flowers attractive to these bats are usually especially designed to encourage bat visits and, for most species, a very wide range of plants are visited through the year and may involve significant shifts in range to follow the fruit and flowering seasons. This mobility is also perhaps responsible for the colonization of many islands around the Indian and western Pacific Oceans, particularly by the larger flying foxes of the genus *Pteropus*. The fruit and flower feeders of the family Phyllostomidae are similar in many of the specialized characters of the Pteropodidae and so show an interesting convergent evolution, but they do retain a proficient echolocation system and long-range movements are only really found in a few of the nectar-feeding specialists, and most of those

at the edge of their range in the subtropics. Species of this family are responsible for the pollination of over 500 plant species in the New World tropics and subtropics, and they also disperse the seeds of at least another 500 species. For a good number of these plants they are the sole or principal pollinator or seed-disperser, and many of these plants are of great importance economically, medicinally or ecologically. The same is, of course, true of the Old World fruit bats. For food plants this includes the oriental durian fruit and the Mexican agaves (used for making tequila), as well as the ancestral stock of such commercial fruits as banana, mango and guava.

In their diet bats show an interesting contrast. The fruit and flowers have evolved characteristics to encourage the attention of the bats for the purposes of pollination and seed dispersal, i.e. the plants try to make life as easy as possible for the bats. On the other hand, particularly amongst the insects, there is an ongoing battle or 'arms race', where the insect prey is developing strategies to avoid being eaten and the bats are developing strategies to keep up with or stay ahead of the insect defences. Here, the insects are trying to make life as difficult as possible for the bats.

The flowers are made easily accessible and with a range of characteristics that enable the bats to find them and feed at them. The flowers may have a lot of anthers and/or nectar, they often only open at night, some are precisely shaped to take the head of a hovering bat or may have a landing platform, they may be coloured (although most are white) such as to encourage the bats to find them, and they may have markings to help guide the bat to the nectar and pollen. Similarly, the fruit is designed to provide smell, taste and texture to attract the bats. In many cases the seeds are included in the part of the plant that the bats eat, and seed dispersal occurs when and where the bats defecate, which may be in flight or at a roost site. With many plants it has been shown that germination is more successful for seeds that have passed through a bat than those that haven't. For larger seeded fruits, the bats will frequently carry the fruit to eat at a distant feeding perch; in this way the seeds are carried away from the parent plant and distributed around the area.

A lesser long-tongued fruit bat, *Macroglossus minimus*, feeding on a banana flower. Bats of this genus are important pollinators of various wild bananas in Southeast Asia.

CHAPTER 4

Breeding and old age

Bats are long-lived animals, with a wide spectrum of species being recorded to live for 30 years or more, but they breed slowly. There are a number of features of bats' breeding systems that are unique amongst mammals.

Breeding

Bats are slow breeders; for the most part they only have one young at a time and only once per year. A few species in the temperate regions, such as the parti-coloured bat and the noctule in Europe, or the hoary bat in North America, more frequently bear twins, as do a number of tropical species ranging from vespertilionids to fruit bats. There is also a number of species that show two birth periods per year, although in some cases this may be different individuals giving birth in different seasons. The most fecund bat is the least pipistrelle, *Pipistrellus tenuis*, of South and Southeast Asia, which may give birth to twins up to three times per year.

Temperate hibernating bats generally have a fairly uniform life cycle. Most of the mating occurs in the autumn, with the male defending a territory from which he makes social calls to attract females, maybe several at a time, for mating. The females may move on once mated,

OPPOSITE: An Australian grey-headed flying fox, *Pteropus poliocephalus*, with its youngster.

RIGHT: A male noctule bat, *Nyctalus noctula*, will call from his roost site to proclaim his territory and attract females. He may also undertake short local song-flights.

The annual cycle of a typical temperate insectivorous hibernating bat.

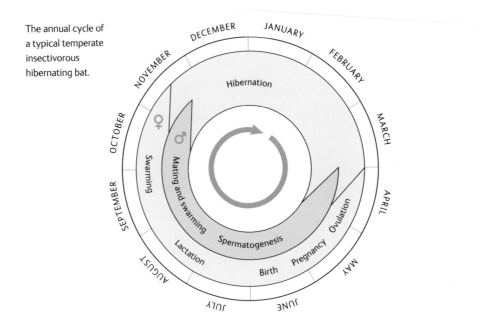

sometimes to another male, and be replaced by others. Thus the mating system is very promiscuous. The males may lose as much as one-third of their weight during this mating period, which leaves them little time in the autumn to fatten up before hibernation. There may be some further mating attempts during hibernation, but little in the spring when the bats emerge from hibernation in relatively poor condition at a time when food resources may be unreliable.

During the winter the females store the sperm and fertilize in the spring. Some other mammals, such as badgers, fertilize straight after mating but delay implantation of the embryo; the delayed fertilization in bats is unique in mammals, although a fairly common practice in insects. Following the spring fertilization and with increasing food availability, pregnancy lasts about six weeks, during which the bat may increase its weight by one-third or more. If conditions are poor they can go into torpor during this period, but that delays the birth period – the earlier the young reaches independence, the higher are its chances of surviving its first winter. As pregnancy progresses the animals tend to collect into maternity colonies, which comprise mainly pregnant females but, depending on species, may include a proportion of non-breeding (immature) females and some males. There may also be non-breeding adults in the colony, usually animals that have not bred this year due to failure of fertilization, or more likely from resorption or abortion of the embryo following poor conditions during pregnancy. The maternity colonies can be quite small, maybe 10 to 20, but in other species, particularly some of the cave-dwelling species, can include up to many thousands of individuals, and some colonies (notably of Mexican free-tailed bat) can be in the millions. The mother suckles her young for about six weeks. Many young bats learn to fly within the roost, although for many bats that live in small tree-holes or crevices around rocks and buildings, their first flights are outside the roost site.

It is widely thought that the mothers teach the young how and where to feed, as well as the locations of alternative roosts, including for hibernacula (sites of hibernation). However, there is a growing body of evidence that suggests that, although the mother has invested a great deal in getting her offspring so far, a few days after its first flight she will leave the young to its own devices. This may reduce the pressure on the food resources around the roost site and allow the female to seek a mate. Or she may wish to start on a migration to a winter quarter remote from the maternity site. Thus the young must quickly develop its echolocation skills, learn to find its own food and feeding grounds, as well as alternative roost sites, and also possibly embark on a long migration.

Baby Mexican free-tailed bats, *Tadarida brasiliensis*, form a crèche while the mothers are out hunting. On its return each mother can find its own youngster.

Another feature of this late summer and early autumn period for many species is a behaviour pattern called 'autumn swarming'. Bats of a wide range of species collect in large numbers at certain places, often underground sites (caves and mines) that may also be important hibernacula, arriving late at night and milling around until the early hours of the morning when they disappear again and all goes quiet. Quite what is going on is not well understood. In some cases it is thought to be related to mating behaviour, and this is clearly true of some sites, but for many it is mainly young bats and males that turn up, which does not suggest that mating is the prime function. Some other species will swarm in other large enclosed spaces, such as the interior of churches, which may also be significant hibernation sites but where, again, the sex and age structure of the participating bats does not always suggest a mating function for the behaviour.

Even in the temperate regions there are variations in the annual cycle. The migration of hoary bats in discussed in Chpater 8 Migration, but once they reach their summering areas they are not colonial like most bats, but almost always roost individually, just hanging in vegetation.

In the tropics and subtropics there is naturally a huge variation to this annual life cycle. There may still be seasonal movement between roosts, but more related to shifting food resources than to the breeding cycle. Both sexes will more frequently roost together so that maternity colonies are more likely to include a higher proportion of males than in many temperate maternity colonies. Other mating and parental strategies are evident. Thus, the males of the West African hammer-headed fruit bat, *Hypsignathus monstrosus*, have a mating system similar to a lek, where a number of males are stationed along the side of a stream, vying with each other for the attention of females that are attracted to their strange honking or croaking sound. To produce this sound, the male exhibits a strangely modified face resulting from a pair of large inflatable air sacs that open into the nasopharynx (and give the

A male hammer-headed fruit bat, *Hypsignathus monstrosus*, calls from the riverside forest of West Africa to attract females.

Bats like this African Chapin's free-tailed bat, *Chaerephon chapini*, probably use their remarkable crest to disperse pheromones to attract mates.

bat its common name), and a greatly enlarged voice box (the larynx) and vocal chords. Some northern species such as the noctule bat and Nathusius's pipistrelle also have a system in which males are loosely grouped together in mating roosts and compete to attract females.

Another system to attract mates, more widely used in the topics, is the use of pheromones (behaviour-altering chemicals) released from glands, usually on the head or wings, particularly in the free-tailed bats and sheath-tailed bats.

In bat colonies there is often a harem structure, in which a dominant male tries to maintain a group of females. Perhaps the best-studied examples are in the spear-nosed bats, especially in the genus *Phyllostomus* (such as *P. hastatus* and *discolor*, which will try to maintain harems of up to 30 females, often in a hole in the ceiling of a cave). Of course, there are always less dominant males lurking around and, despite the best efforts of the dominant animal, they may be responsible for a significant proportion of the mating. The common vampire bat, several sheath-tailed bats and some fruit bats maintain well-mixed-sex colonies throughout the year.

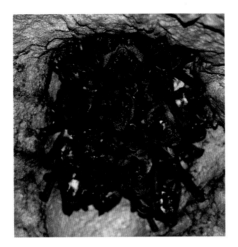

ABOVE: A harem of greater spear-nosed bats, *Phyllostomus hastatus*, in a solution hole in the roof of a cave, Trinidad.

RIGHT: A pair of yellow-winged bats, *Lavia frons*, roost in a tree in the African savannah.

In the mixed-sex colonies or where bats use seasonal mating roosts, they are generally regarded as promiscuous in their mating. However, it is clear that in a number of species in which DNA has been used to establish parentage, certain males are achieving more than the equal share of mating (as in the little brown bat in North America or the greater horseshoe bat in Europe). Apart from the harem groups, where one male is trying to maintain exclusive management of several females (i.e. where mating is less promiscuous), there is also quite a broad range of mainly tropical species that appear to be monogamous, from the large carnivorous Neotropical false vampire bat, through some Old World false vampire bats (Megadermatidae), slit-faced bats, woolly bats, even some horseshoe and leaf-nosed bats and at least one species of flying fox.

While delayed fertilization is the norm for temperate hibernating bats, at least the European bent-winged bat exercises delayed implantation. In some parts of its range, this species (and forms now separated as sister species which may also demonstrate delayed implantation), also have another mechanism to add flexibility to their reproductive strategy. This is a further delay in development after implantation. This delayed development is the least common form of reproductive delay and is not recorded in other mammals, but it is recorded in a few other tropical or semi-tropical bat species, such as the California leaf-nosed bat, *Macrotus californicus*. One further resort available to temperate bats, as mentioned earlier, is torpor in response to poor conditions before birth. This can delay birth, but there is a risk in such a strategy, in that it is important to give birth early to ensure that the young is flying and independent as soon as possible, to give it a better chance of survival through its first year. It may also disrupt the synchrony of births in colonial bats. As with the temperate bats, colony size in the tropics can vary up to several million individuals, most of the larger colonies being of free-tailed bats or fruit bats. The colonies of species from several groups, including leaf-nosed bats, horseshoe bats and bent-winged bats, can number in the hundreds of thousands.

The young

The baby bat, often called a pup, is born relatively large (sometimes more than one-quarter the weight of the adult), but is naked, with eyes closed and not able to regulate its body temperature; this is described as being in an altricial

Mexican free-tailed bats, *Tadarida brasiliensis*, emerging from a cave in southern USA. Colonies of up to 20 million have been recorded.

state. The newborn pup quickly attaches to a nipple or is guided to one. It may stay on the nipple for most of the day and in some species (mainly fruit and flower feeders) may be carried by the mother on foraging trips for an initial period; however, this is more the exception than the rule – or it happens by mistake. For the most part, the young are left behind in a crèche when the mother is foraging away from the roost. The mammary glands of the mother are relatively large and may make up 10 per cent of the body weight. For almost all species there is one pair of thoracic nipples, but in a very few, such as *Lasiurus*, *Vespertilio* and *Otonycteris* species, which regularly have two and occasionally more than two young, there are four functional thoracic nipples. Members of about five families of bats, including the horseshoe and Old World leaf-nosed bats, have a pair of 'false nipples' in the pelvic region, which, although capable of producing some milk, mainly act as 'holdfasts' for the youngster to attach to. The infants do spend a lot of time attached to these pubic nipples and with their legs wrapped round the mother's neck. The males normally have a pair of rudimentary nipples, but in a very few species, notably the Dyak fruit bat (*Dyacopterus spadiceus*), the male has developed thoracic glands from

A Mehelyi's horseshoe bat, *Rhinolophus mehelyi*, flies with its baby attached to its pubic nipple.

which some fluid can be expressed; this expression of milky liquid is known as galactorrhoea and may result from the consumption of plants rich in the female hormone oestrogen.

For the most part the females only suckle their own young, and even in the largest colonies the mothers have a range of cues to enable them to find their own youngster. The only example of a pup being fed anything other than milk is with common vampire bats, which may feed an older youngster, and not necessarily its own, with regurgitated blood.

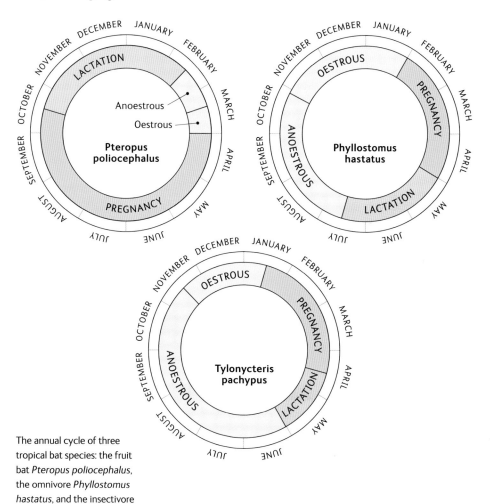

The annual cycle of three tropical bat species: the fruit bat *Pteropus poliocephalus*, the omnivore *Phyllostomus hastatus*, and the insectivore *Tylonycteris pachypus*.

Where they have the opportunity, the young may start to fly well before they are weaned, but weaning will not occur until they have developed about 80 per cent of the wingspan of a full-grown bat.

Ageing

For the most part, bats are born with a sex ratio of about 1:1. As we have already seen, the ratio of male to female in breeding colonies can vary widely, with males being almost absent from many colonies. On the other hand, bats at autumn swarming sites and in underground hibernation sites show a heavy bias towards males. Thus, for these and other reasons, it may be difficult to establish the true sex ratio of field populations.

Bats are very long-lived. For a whole range of species, from large fruit bats to small insectivorous bats, it has been shown that individuals can live to over 30 years. The current record is 41 years, held by a Brandt's bat, *Myotis brandtii*, a species that occurs across the Old World temperate region and weighs about 6 g (0.2 oz), with this individual ringed in Siberia. Of course, relatively few bats will attain these great ages. It is likely that for most species there is a survival

Brandt's bat, *Myotis brandtii*; this species is the holder of the longevity record for bats.

A red-tailed hawk takes a bat from a stream of Mexican free-tailed bats, *Tadarida brasiliensis*, emerging from a cave in Texas.

rate of about 50 per cent in their first winter. Once mature the survival rate may increase to about 70 per cent per year. Probably for most species, the average life expectancy is about five to seven years, but even this is a very long lifespan for a small animal with a very high metabolic rate.

Females of the European common pipistrelle may mate in their first autumn (at about three months), but most bats do not reach sexual maturity until they are one or two years old, and in some cases as much as five years or more. Even then they may not give birth every year, especially in their early years; indeed, older animals may be more reliable breeders. There is no evidence that the fertility of the females drops off with extended age – even the oldest greater horseshoe bats, at about 30 years, have been shown to be still breeding – and the young of these older females seem to have a higher-than-average survival rate.

As we have seen, bats are *K*-strategists, which means that they breed slowly and hence live longer in order to produce enough young to maintain the population – most animals of their size have the opposite strategy of 'live fast, die young' (known as *r*-strategists). A range of factors contribute to the ability of bats to live a long time. Bats are subject to relatively low levels of predation. There are so-called bat hawks in the tropics; as with some other birds of prey, these wait at dusk for the emerging stream of bats from large colonies and pick off individuals from the crowd, but the number of bats taken is relatively small and unlikely to have significant impact on the population. Similarly, owls and even more generalist feeders, such as magpies, will learn to pick bats off as they emerge from particular roost sites, which can be a problem for some rare and vulnerable populations. One surprising recent observation in Europe was of a great tit, *Parus major*, repeatedly taking and eating bats that were in hibernation near the entrance to a cave. Mammals such as stone martens and rats have also been recorded taking hibernating bats. And there are a few large carnivorous bat species, such as the Central and South American greater spear-nosed bat (*Vampyrum spectrum*), the African large slit-faced bat (*Nycteris grandis*) and the Australian ghost bat (*Macroderma gigas*) that sometimes take smaller bats. Even the largest European species, the greater noctule (*Noctilio lasiopterus*) has recently been shown to feed on small migrant passerine birds and occasionally other bats. In the tropics, there are frogs and snakes and other animals that take bats while roosting or as they emerge from the roost.

None of these examples of bat predation is likely to have a major impact on the population (a notable exception of a snake and the Guam fruit bat (*Pteropus tokudae*) will be dealt with later; see p.90). Similarly bats are host to a wide range of parasites (see Chapter 7, Predators), but which seem to have relatively minor impact on the bats themselves, unless perhaps the bat is in poor condition for other reasons. Neither is disease a widespread problem. There have been reports of mass mortalities, particularly of *Miniopterus* species in Europe, Africa and Australia, but the cause of such events has not been identified, although there is strong inference that it has been caused by a filovirus, Lloviu virus.

One disease which has caused huge mortality that has threatened the populations of some species is white-nose syndrome (WNS) in North American bats. This is caused by a fungus, *Pseudogymnoascus destructans*, which invades the tissue of the bats while they are hibernating in cold caves and mines. It can be particularly obvious as a white dusting around the nose (hence the name) and elsewhere on the face and ears and on the wing membranes. Associated with these symptoms is irritation caused by the fungus, which results in arousal from hibernation and hence numbers of bats flying around the entrance to hibernation sites, and mass mortality because the arousals burn up so much fat that the bats starve to death. First reported in 2006 in the Northeastern USA, it rapidly spread, particularly down the eastern side of the USA, and has now been recorded from at least 33 US states and seven Canadian provinces. It has killed millions of bats. The fungus was searched for in Europe and quickly found, first in France in 2009, but since then widely through Europe and Russia and into China. But, although the fungi found here are inseparable from those in North America, here there are none of the other symptoms of WNS. It is likely that the fungus has been around in Europe for a very long time and the bats can live with it, while it may be a quite recent introduction to North America, where there was no immunity. It is further likely that it was introduced by a cave visitor, such as a caver, bat worker or cave tourist, and it is perhaps a classic example of the risks to wildlife of our global society.

Bats also die of other diseases, such as rabies and Hendra virus, but not in vast numbers. They are more likely to die through factors such as having an accident, from starvation or, perhaps more frequently, from lack of water.

The exceptional longevity of bats has aroused interest in the medical world. One project from Emma Teeling and her team at University College Dublin has investigated telomeres, which are protective caps of nucleotide repeats on the tips of each strand of DNA; the telomeres act like the plastic tips at the end of shoelaces to protect the chromosomes. The telomeres shorten with repeated cell division (the ends become frayed) until they are so short that the cells stop dividing, causing the age-related breakdown of cells and potentially limiting lifespan. Cells have telomerase, an enzyme that can restore telomere length and thus prevent the cells from growing old. The project showed that telomeres shorten with age in some bat species (e.g. *Rhinolophus ferrumequinum* and *Miniopterus schreibersii*), but not in others, including in the genus of bats with the longest recorded lifespan, *Myotis*. The project showed that, rather than telomerase, an array of telomere maintenance genes that repair and prevent DNA damage potentially mediate telomere dynamics in the *Myotis* bats. The project demonstrates how telomeres, telomerase and DNA repair genes have contributed to the evolution of exceptional longevity in *Myotis* bats at least.

CHAPTER 5

Roosts

A roost or roost site is here considered to be a place where a bat rests between flying periods. Probably the most widespread roost sites are caves and trees, but a large number of bat species now also roost in buildings and other artificial structures. A wide range of bats, especially in the tropics, also roost in vegetation, and some modify the vegetation to suit their needs.

From roosts in caves and tree holes....

Caves are traditional roost sites for a large number of bat species and are where most of the larger aggregations can be found. Natural caves provide a wide range of conditions for the bats to choose from and so they can provide for a wide range of species that use the caves for a wide range of functions. Natural lava tubes also provide a similar habitat, and over the last few millennia the digging of mines and other underground passages has added to the opportunities for bats to roost, especially in areas that were otherwise deficit in natural underground features. This includes the construction of underground or semi-underground structures, such as fortifications, 'secret' tunnels, and smaller structures, such as culverts, cellars, ice houses, kilns and food stores. Caves and similar artificial underground spaces can provide a range of temperature and humidity, a complex structure that provides exposed and more-or-less flat roosting areas as well as cavities and crevices, and offers protection from most predators. The size and configuration of any entrance is also important in controlling the air flow and the extent to which warm or cold air is retained within the system. In temperate Europe, such places are mostly used by hibernating bats and provide a constantly cool environment,

A cluster of greater horseshoe bats, *Rhinolophus ferrumequinum*, hibernating in a cave in France.

A whiskered bat, *Myotis mystacinus*, hibernating in a tunnel. The low temperature of the bat has resulted in condensation settling on its fur.

but warmer sites are also used in summer, especially by horseshoe and bent-winged bats, and some *Myotis* species.

There are many records of bats being found hibernating in the scree on the floor of caves and mines (mainly *Myotis* species) and in natural mountain scree (*Myotis* and *Eptesicus* species). Other records of bats roosting at ground level are uncommon, but there are several records of *Murina* (a genus of the vesper bat) being found hibernating on the ground beneath the snow in Japan and the Russian Far Eastern Federal District.

Cavities in trees are the other main natural roost site for a wide range of bat species. The cavities may be natural rot holes, or holes created by other animals such as woodpeckers and barbets or squirrels,

A group of Schreibers' bent-winged bats, *Miniopterus schreibersii*, densely packed on the wall of a cave.

and we can include here the cavities created by loose exfoliating bark and by breaks or damage to the bark. Sometimes the tree bark is sufficiently fluted as to provide suitable crevices for bats to roost, or encrusting vegetation such as ivy may provide cover for roosting bats. The cavities in trees will never match the space and diversity of a roost site offered by a cave or similar structure, but they do offer sites with some natural heat produced by the timber itself and with some natural insulation from extremes of temperature, and some protection from predators. Perhaps because of the rather different conditions, the species using trees cavities in summer are generally quite different from those using caves. Tree holes are more regularly used in the temperate regions by species of *Pipistrellus*, *Nyctalus*, *Plecotus* and *Myotis*, although even in the north temperate regions species such as Daubenton's bat, *Myotis daubentonii*, may use either habitat for its maternity colonies.

There is a tendency for tree-roosting colonies to change roost site frequently such that they know a number of available roost sites in the vicinity. The reasons for this are not clear; it has been suggested that it is to prevent parasite build-up, but a very high and diverse parasite load does not seem to affect many

Noctule bats, *Nyctalus noctula*, generally choose tree holes for their summer roosts and may also hibernate in tree holes.

cave bat species, and even some tree-dwelling species tolerate the attention of a range of specialist parasites. The colony movement of Bechstein's bat, *Myotis bechsteinii*, in Europe has been studied in detail by Gerald Kerth and his team at the University of Greifswald, Germany. With ringing and DNA studies they have been able to identify neighbouring maternity colonies of the bats as being quite discrete, demographically independent units, with mating outside the territory providing gene flow. While there is high colony fidelity and very little movement between colonies – indeed, evidence of resistance to immigration – the constituent members of the colony may include several distinct matrilines. Thus, these colonies maintain their separate territories and there is very little exchange between colonies. Within the separate 'mother' colony, the population is frequently breaking up into smaller groups, which then regroup to a varying extent (described as a strong fission/fusion society), so that the colony is continually on the move within a fairly limited area of 15–40 ha (37–100 acres), and an individual bat is generally moving roost site every two or three days.

A recent discovery in Poland has been the use by Nathusius's pipistrelle bats, *Pipistrellus nathusii*, of holes in tree boles left by the emergence of the longhorn beetle, *Cerambyx cerdo*. The bats use the holes left by the emerging beetles as a hibernation site and are found 50–150 mm (2–6 in) from the edge of the exit hole. They were mainly between 1.5 m and 3 m (5–10 ft) above ground level, in holes facing south to southwest, and showed a preference for oak trees with a high abundance of *C. cerdo* cavities.

...to buildings

Perhaps bats have lived with us ever since we started to construct permanent homes, but it is certain now that with the huge losses in the availability of suitable trees, and the damage and destruction to many caves, bats have followed the wood and the stone into buildings, where we have replicated some of their favourite roosting places. The older, larger buildings, such as castles and temples, cater for a wider bat diversity than the newer buildings of simple construction that seem to be favoured mostly by the more common and adaptable species, such as pipistrelle bats and *Eptesicus* species. Bats can also be separated into the crevice dwellers – such as the pipistrelle bats and some *Myotis* species, which probably mainly originated as tree dwellers and now live in gaps in big old timbers or behind wall claddings such as timber, tiles or manufactured sheeting – and the

Some *Eptesicus* species, such as these European serotine bats, *E. serotinus*, or North American big brown bat, *E. fuscus*, generally form their colonies in buildings.

cave or rock dwellers, such as the horseshoe bats and *Eptesicus* species that now roost in open roof spaces. Here they find the warmth and shelter they need for breeding sites, and those that need it might find a basement cellar within which to hibernate. The larger structures offer a wide range of available roost situations such that, without changing roost site, the bats can choose their ideal conditions for the time of year or stage within the breeding cycle, or reflect the effect of the ambient weather conditions. Throughout the world there are house bats that dominate the occupation of buildings, including free-tailed bats throughout, and species of *Carollia* and *Glossophaga* in Central and South America, *Scotophilus* and *Hipposideros* in Africa and Asia, and species of *Chalinolobus*, *Scotorepens* and *Vespadelus* in Australasia.

Roosts in vegetation

Although most bats need shelter for their day roosts, a number of bats roost out in the open. Probably only the larger fruit bats choose to roost in tree canopies,

LEFT: A camp of little red flying foxes, *Pteropus scapulatus*, roosting amongst the foliage of an Australian tree.

ABOVE: A group of African straw-coloured fruit bat, *Eidolon helvum*, cluster in a tree.

where they are exposed to direct sunlight and where they form very large 'camps'. The larger fruit bat camps, of most of the *Pteropus* and *Acerodon* species of islands in the Indian and Pacific oceans and *Eidolon* of the African mainland, are noisy and conspicuous. These bats generally use the most emergent trees, and their concentrations may cause defoliation. An important feature is that the bats need to be able to drop off their roost site into free flight, and so the size and manoeuvrability of the bat will to some extent dictate the roost site in vegetation.

A number of sheath-tailed bats, such as *Emballonura* in Africa and *Rhynchonycteris* and *Saccopteryx* in Central and South America, roost in exposed situations on tree trunks or on artificial structures such as bridges or buildings, but usually on the shady side. Particularly *Emballonura* will usually be close to some vegetation that provides extra shelter. As its name suggests, the African yellow bat, *Lavia frons*, is bright yellow, but it roosts just hanging in a bush, tree or rock overhang, where it is surprisingly inconspicuous (or at least is passed unrecognized).

Many other bats roost in the hanging dead foliage of the tree itself or its epiphytes. Even in the temperate regions, species of *Lasiurus* roost attached to the leaves of densely foliated canopy trees, often in or on suspended clumps of dead leaves, as does *Murina* in Australia. Such roosts tend to be used by individual or small groups of bats and are often quite temporary, with the bats frequently changing roost site.

The furled leaves of certain plants such as banana and, in the new World, *Heliconia* and *Calathea* are favoured as roost sites. Specialists for this type of roost site, such as *Thyroptera* in the New World (Central and South America) and

LEFT: The North American hoary bat, *Lasiurus cinereus*, usually roosts singly or in small family groups amongst the foliage of woodland trees.

BELOW: A group of disc-winged bats, *Thyroptera tricolor*, roost attached by the pads on their thumbs and feet to the inner wall of a *Heliconia* leaf. They roost head-up.

Myzopoda in Madagascar have well-developed pads or suckers on their wrists and feet to help them hang to the smooth vertical wall of the leaf. Not so well adapted are species such as *Pipistrellus nanus* in Africa and one or two species of *Myotis* in Asia. These, too, are bats that tend to roost individually or in small groups of rarely up to ten bats. That they do live in such small groups and in such temporary, even ephemeral, roost sites (a leaf normally unfurls enough to become unusable after one to three days) does have one advantage – the bats are spared the attention of a wide range of ectoparasites that can affect the inhabitants of large colonies in more permanent roost sites.

For the most part, there is no effort by bats to create a roost – only to use what is there. However, there are bats that do modify natural structures to create or enhance a roost site. Most of these are the tent-making bats, which create a sheltered roost area from natural vegetation. Mostly these bats modify individual leaves by chewing the main leaf ribs so that parts of the leaf fold down to create a sheltered area. So these 'tent-makers' also live in somewhat temporary accommodation. Of 23 species that have been recorded as constructing or roosting in such tents, 18 belong to the stenodermatine fruit-eating spear-nosed bats of Central and South America, such as species in the genera *Ectophylla*, *Artibeus* and *Uroderma*. The way these bats modify the leaves depends on the shape of the leaf, so that about eight basic styles have been identified. With a fan-shaped palm leaf, the bat chews the main ribs a few centimetres from where they join the stem so that the rays fold downwards and the leaf looks rather like a badly collapsed umbrella. With elongate leaves of such plants as *Musa* (banana) and *Heliconia*, the bats chew the ribs diverging from the central midrib, such that again the outer parts of the leaf fold down to form the tent. Other leaves, particularly of palms and Araceae (a rather large and diverse family that includes the arums) are approached in different ways but with the same overall effect. Less frequently, a group of leaves may be attacked to make a similar canopy. Most of the core group of tent-building species will have their own preferred host plants and means of modification, but some individual species do have a broad range of host plants and a range of architectural styles to fit the range of plants. Finding these tents can be very difficult at first, but having developed a 'search image', leaf tents can be seen to be very common in certain, mainly forest, areas. These tents have a number of advantages in protection from inclement weather and direct sunlight, from predators hunting by vision, the ability to construct roosts close to feeding territories, and the avoidance of parasites. There may also be some

The Central American Honduran white bat, *Ectophylla alba*, roosts in small
groups under tents constructed from the leaves of forest trees such as
palms and *Heliconia*.

energy-saving advantages in living under one of these tents, but that might have
to be set against the energy costs involved in tent construction; the leaf must be
chosen carefully, and while some of the simpler tents may be constructed in one
night, others may take several nights.

These tents may be constructed by male bats which then attract a group
of females, although in at least the Honduran white bat, *Ectophylla alba*, both
sexes cooperate in the tent-building. They mostly choose younger leaves, which
probably have more flexibility and are easier to 'work', but still the roost is fairly
temporary and is likely to become unusable after a few weeks, rarely lasting up
to six weeks. Tents may be occupied alternately by more than one species. One
outlier from the main group of tent-builders, *Rhinophylla pumilio*, may not make
its own tents, but occupies the abandoned roosts of other bats.

Guilherme Garbino and Valeria da Cunha Tavares from Brazil have recently
used DNA, and a technique called stochastic character mapping, to investigate
the origins of foliage roosting in stenodermatine spear-nosed bats. They found
records of 48 species roosting in foliage (including standing or fallen tree trunks),
with tents used by 19 species, and that foliage-roosting was monophyletic (i.e.

all these bats developed from a common ancestor) and developed early in their evolution. It combined the Stenodermatinae and Rhinophyllinae as one sister group, and was closely linked to frugivory. But within that group they found at least two independent developments of tent roosting. The tent-dwelling bats tend to be smaller species and to have distinct bright facial and/or dorsal stripes, or shoulder patches. The tent-dwellers also tend to live in small groups (occasionally up to 20); this might be expected considering the weight-bearing capacity of the leaves, but the authors suggest that other factors, such as phylogeny, may also play an important role in interspecific differences in group size.

In the Old World tropics three small species of fruit bat are also tent dwellers: *Balionycteris maculata*, *Cynopterus brachyotis* and *C. sphinx*. The latter two species build tents that are remarkably similar to three of the designs used in the spear-nosed bats, but all three species also construct so-called stem tents, which they cut into the root mass at the base of dense trailing epiphytic plants such as ginger or bird's nest fern (*Balantiopteryx*) or similar growth of other plants, including the palm *Caryota* (*Cynopterus*). These stem tents can be substantial works and may take months to complete, and in the case of these fruit bats it does seem to be the male that does all the work and who then encourages females to join him.

There is one insectivorous tent dweller. In the Philippines one of the lesser house bats, *Scotophilus kuhlii*, occupies tents in one of the palms also used by *Cynopterus* species. Small groups will use these umbrella-type tents, but it is not clear whether they build their own tents or occupy those built by the fruit bats.

The genus *Kerivoula*, the woolly bats, includes some unusual choices of roost site. Some species regularly roost as isolated individuals or small groups in clumps of dead or dying vegetation, but in Africa and South Asia species of this genus have also been found roosting in old weaver bird nests, even in a weaver bird nest with an active wasps' nest in it. A number of other bat species also use birds' nests as a roost site, including *Murina* in Australia and Southeast Asia. These bats choose suspended nests (similar to weaver bird nests) and are mostly unmodified, but it would seem that *K. papuensis* usually ensures that there is a basal hole in the nest for viewing and ease of access. Mud nests of birds, especially the more tubular nests of some hirundines are used by several bat species, including *Tadarida brasiliensis*, but usually only as individuals and mostly rather erratically or opportunistically.

Ant or termite nests are also used by bats, sometimes ground nests in Africa by *Nycteris*, sometimes aerial nests in Central/South America by *Lophostoma* following a male bat excavating a hole in the nest from below.

UNUSUAL SITES

There are beetles that lay their eggs in the younger stems of bamboo, particularly in Southeast Asia. Eventually the adult beetle emerges, creating a small hole in an internode (the area between horizontal divides, or nodes) of the plant stem. As the bamboo continues to grow, until it hardens the hole gets a bit stretched and may end up c.20 mm (¾ in) long and up to 10 mm (½ in) wide. That is ideal to provide access for the bamboo bats. Species of *Tylonycteris* are the smallest bats in the world and can weigh as little as 2.5 g (just under 1 oz), and they have a very flattened head. They live individually or in a small group within the internode of the bamboo culm, and, despite their size, will defend aggressively their roost territory. Group sizes are usually small, but over 30 in one cavity has been recorded. These bats have thickened pads on their wrists and feet to help hang onto the smooth vertical inner surface of the bamboo. *Eudiscopus denticulatus* and *Glischropus tylopus* also do the same, although the latter does not have quite the same flattening of the skull. As the bamboo deteriorates and the hole gets worn, other bats begin to gain access and a range of species is recorded, especially the smaller pipistrelle species such as *P. mimus*.

Another unusual roost site for a *Kerivoula* species is that of *K. hardwickii* within the 'pitcher' of a pitcher plant, *Nepenthes rafflesiana*; here the bats just fit within the pitcher and can hang onto its inside with their feet. If they were to slip down into the bottom of the pitcher they would die and ultimately be dissolved to feed the plant, but the pitcher gets too narrow for them to slip down as low as the liquid level. The plant provides protection for the bats, and the bats feed the plant by depositing their droppings into the pitcher.

LEFT: A lesser bamboo bat, *Tylonycteris pachypus*, emerges from a bamboo stem.

ABOVE: Hardwicke's woolly bat, *Kerivoula hardwickii*, roosting at the lip of the pitcher of *Nepenthes rafflesiana*.

Roost function: maternity and hibernation

The roost site might serve any of a number of functions, such as for maternity colonies, as mating sites, as sites for resting during the night or between feeding bouts (which ornithologists may refer to more as 'loafing' rather than 'roosting'), or as temporary transient or gathering roosts where bats accumulate over a relatively short period before moving into a more seasonal roost, such as those used for maternity or hibernation. For some bats, larger colonies will gather and then break up temporarily into smaller groups, reassemble and break up again throughout the season; this is referred to as a fission/fusion society. In almost all bats there are seasonal shifts in roost sites in response to changes in weather and food availability. The practice of spring and autumn swarming is discussed elsewhere (see p.52), but roosting is not the main function of the site for this activity, and so these are just referred to as swarming sites, although many are recognized as roosts for other purposes.

The two key roosts are maternity roosts and hibernation roosts. Maternity roosts are sites where breeding females gather together as a colony to have their young; a 'colony' is a group of animals that have gathered together for a particular purpose, usually breeding. They may be joined by a variable proportion of males depending on the species, the nature of the colony (such as whether it is matriarchal or harem structured) or how far they are through the pregnancy and lactation period (males often leave the maternity colony at about the time of birth of the young). Maternity sites are usually quite warm to prevent too much heat loss by mothers and babies. Where there is not enough intrinsic heat, the colony may be of sufficient size and carefully situated so as to create its own heat. Some species, such as the ghost-faced bats (*Mormoops* species), deliberately seek so-called 'hot caves', which have a high natural internal temperature and humidity. And some colonies seem to be able to tolerate, if not welcome, extremes of heat and perhaps desiccation, such as beneath a tropical corrugated iron roof, as with many of the free-tailed bats.

In temperate regions bats hibernate in winter when insect food availability declines. Hibernation sites need to be relatively cold, but generally not fall below freezing, and need a high humidity for most species. Here bats can drop their body temperature, heart rate and all metabolic functions to a very low rate. For example, an actively flying bat may have a heart rate of up to 200 beats per minute; at rest this might drop to *c.*70 beats per minute; and while in deep hibernation it

Colonies of Mexican funnel-eared bat, *Natalus stramineus,* roost in warm
caves but roost neatly spaced apart – each with its own personal space.

may drop to 15 beats per minute. In this condition, their body temperature may
be very close to that of the surroundings. If they need to wake up it may take up
to 30 minutes before they are able to take flight, despite the fact that they can
metabolize exothermic brown fat, found mainly around the shoulders, to assist
the arousal process. Although some bats may hibernate for extended periods,
mostly they maintain a basic diel (daily) rhythm of increased metabolic activity
at around the normal emergence time, and many species arouse quite frequently
through the winter to excrete and drink, and, when the opportunity presents
itself, to feed. Natural arousals can be accommodated in the fat reserves and
ancillary feeding, but an unscheduled arousal can cost a considerable amount
of the stored energy, perhaps as much as enough for ten days' hibernation. Bats
may roost individually or in clusters (which may be of mixed species), but they
choose their particular roost site depending on the species, the time of winter,
the ambient conditions, their individual age, sex and condition. Thus there is a
tendency for long-eared bats to roost close to the cave entrance, where it may
be colder, so that they can drop their body temperature further to save energy;

here they risk being caught out by a deep freeze and forced to move roost site, but it is also where they can take advantage of warmer winter spells to forage for insect food if the outside temperature rises above the threshold for insect flight. Deeper in a cave system the temperature may be warmer and more stable, thus costing a little bit more energy for a Daubenton's bat to roost at the surrounding temperature, but allowing longer periods of uninterrupted sleep through not being so affected by changes in ambient conditions. But even the Daubenton's bat, as the winter progresses and if its stored reserves get dangerously low, may try to find a colder site to enable it to eke out its energy reserves for that final period.

Bats may collect from a considerable area, including after lengthy migrations, to a favoured hibernation site. Traditionally these would have been cave sites or other rock crevices and tree hollows. Perhaps even scree is more important than has been realized, but it should be remembered that bats are extremely vulnerable to predation in hibernation and so are safest well clear of the ground.

Other roost functions

In between the hibernation and maternity roosts temperate bats need a range of other roosts. Most of these roosts are used rather temporarily, such as the transient gathering roosts used particularly in late spring when bats start to aggregate into their summer colony, but once the bulk of the summer colony has grouped together they move on to their maternity site. Perhaps even more temporary are the stop-over roosts of bats on migration, but as bats migrate in short hops, certainly the long-distance migrants need a large number of such roosts along their migration route. During the summer, many bats use night roosts away from the main colony site. Here they will rest for periods between feeding bouts before returning to the main day roost in the morning. Such night roosts may also double up as feeding roosts for species such as the long-eared bats or horseshoe bats. These species regularly take relatively large prey, such as beetles and moths, and carry them to a regular site, where they dismember the prey, dropping inedible material to the ground and eating the rest. They may return to such sites through the night and over long periods, such that a good indication of the diet and the seasonal changes in diet can be seen from the prey remains below the roost. Normally the females leave their baby in a crèche at the main summer maternity site, but a few species, such as the Indian false vampire bat, *Megaderma lyra*, leave their babies in a separate roost while the mothers are off hunting during the night. Following the maternity period, there is the mating period, when males may hold territories and defend a roost site in a cave or tree hole and invite females for mating; sometimes these roosts are individual or sometimes for a group of males. This pattern of changes in roost use through the season is an important factor in the conservation of such bats.

For many of the sedentary tropical species, there may be no need to change roost sites, and they may use the same roost site all year – and for many years. Shifts in roost sites in tropical and subtropical bat species are more associated with changes in foraging opportunities for both insectivorous and fruit- and flower-feeding bats – particularly the latter. Evidence for the sedentary nature of some bat colonies can be seen in the huge piles of guano that can be seen below the regular sites of large bat colonies. And their permanence is also apparent in the range of especially evolved invertebrates and other fauna associated with these guano piles. The human exploitation of such guano accumulations for fertilizer is a long-established practice in some areas, and we will return to that topic (see p.129).

A lone eastern red bat, *Lasiurus borealis*, hangs in a
North American pine tree.

CHAPTER 6

Migration

Many bat species undertake regular seasonal migrations. Although they do not go on the great long-range migrations of some birds, movements of up to 2,000 km (1,300 miles) or more are known for a range of species, both temperate and tropical. There have been various attempts to classify migratory bats into long-range, medium-range and short-range (or sedentary) species, based on the maximum distances known for the species. While useful in many ways, such classifications are complicated by migration of one sex more than the other, or only certain populations of a species being migratory. For some, the movements are not regularly between point A and point B, but rather the route and distance may vary with differing flowering and fruiting patterns between years.

Temperate

In temperate insectivorous bats, migration may be more for switching between breeding sites and suitable hibernation sites; in tropical regions it can be more related to seasonality of food supplies, especially for fruit- and flower-feeding bats. In temperate regions there may be a strong sexual bias, with the males remaining in the wintering areas while the females migrate north to maternity areas, where there is less competition between bats for summer roosts and food but less availability of suitable hibernation sites. Even in the tropics there may be some bias in who moves; thus in the small African fruit bat, *Myonycteris torquata*, it is mainly the immature males that migrate, whereas in the Australian black flying fox, *Pteropus alecto*, it is the young that stay behind when the adults disperse from the breeding colony. In the American hoary bat, *Lasiurus cinereus*, perhaps

Straw-coloured fruit bat, *Eidolon helvum*, is a long range migrant in Africa.

The hoary bat, *Lasiurus cinereus*, migrates from much of North America to Mexico and is thus a long-range intercontinental migrant.

the best example of an intercontinental migrant, both sexes migrate from North America to Mexico (and some may go further south than that); later, however, as they migrate north towards the northern summer, the sexes become geographically separated, with the males ending up in the west, and with mostly females reaching the northern ends of the range in central and eastern USA and Canada.

Almost all insectivorous bats of temperate areas move from a warm summer site suitable for pregnancy and lactation and with adequate food supplies, to a cool winter site for hibernation. This movement might take bats of the same colony from the roof space of a building to its cellar or to a location far from the summer site. Winter aggregations in hibernation sites can include individuals of a single species collected from a wide range of distances. In Europe, particularly for the medium-range migrants, such as the greater mouse-eared bat (*Myotis myotis*) or pond bat (*Myotis dasycneme*), their movements to the hibernation site may be in any direction. The longer-range migrants show a much greater bias for north–south, or rather northeast to southwest migration; nevertheless, many Nathusius's pipistrelle bats, *Pipistrellus nathusii*, moving the 1,500 km (1,000 miles) from Latvia and Lithuania to the UK are moving almost westerly.

Migrations of up to almost 2,000 km (1,300 miles) are also recorded for the Mexican free-tailed bat, *Tadarida brasiliensis*, both in the north and south of its range. Migration is much better studied in the north end of the range, from its natal sites in southern USA to Mexico; little is known of the migration at the southern end of the range in Argentina beyond the seasonal absence of large populations there. Migration is again mainly females moving to higher latitudes to form nursery colonies. Some males also travel north to the maternity sites, and they usually arrive first. Some of the US populations are non-migratory or only undertake short local movements (particularly on the western (Pacific) side and east of eastern Texas) but the populations in between are strongly migratory. Various techniques, including analysis of DNA, have been used to assess whether the migratory and non-migratory populations are distinct forms, but no such separation has yet been justified. Some of these colonies form the largest aggregations of mammals, with up to 20 million bats accumulating in one cave. Some of these caves provide spectacular viewing opportunities for people to watch the evening emergence; equally popular is Congress Avenue Bridge in Austin, Texas, where over a million bats can easily be seen emerging from their roost under the bridge. The migration is sometimes in large flocks, but the bats are thought to move about 50–70 km (30–40 miles) per night between stop-off points, which may be used for one or a few days but are very important for successful migration. Intriguingly, there are three relatively recent records of this species being found alive on the Falkland Islands in circumstances where there is no evidence of assisted passage. This is *c*.2,000 km (*c*.1,300 miles), and almost all across sea, from the nearest part of their natural range in Argentina, and it could be an extreme example of 'overshooting' or of 'reverse migration', which is regularly reported in birds but not in bats.

So we begin to see that there are species that regularly undertake long migrations such as: in Europe Nathusius's pipistrelle (*Pipistrellus nathusii*), the parti-coloured bat (*Vespertilio murinus*) and noctule bat (*Nyctalus noctula*), and in North America species of *Lasiurus* (especially the hoary bat), silver-haired bat (*Lasionycteris noctivagans*) and Brazilian free-tailed bat (*Tadarida brasiliensis*); some middle-range migrants: such as in Europe the greater mouse-eared bat, the pond bat and Schreibers' bent-winged bat (*Miniopterus schreibersii*) and the little brown bat (*Myotis lucifugus*), tricoloured bat (*Perimyotis subflavus*) and *Lasiurus seminolus* in North America; and some species that only undertake very short movements or are rather sedentary: such as most *Rhinolophus* spp, *Plecotus* spp and the serotine bat (*Eptesicus serotinus*) in Europe, and the big brown bat (*Eptesicus fuscus*) and *Corynorhinus* spp in North America.

Tropical

For the tropical fruit bats, it was widely recognized that a number of species are migratory, but details were sparse. The recent introduction of smaller satellite tags and other forms of marking have made studies much more practical. Between five and ten million straw-coloured fruit bats, *Eidolon helvum*, congregate in a 9-hectare (22-acre) patch of swamp woodland in Kasanka National Park, Zambia, for a fairly brief period in November and December. Recent tagging has shown animals returning up to 2,000 km (1,300 miles) into the Democratic Republic of the Congo, but doubtless these bats gather from a wide area to take advantage of a wealth of seasonal fruits in Zambia. In Southeast Asia, the seasonal presence and absence of colonies of the large flying fox, *Pteropus vampyrus*, were used to gain information on its migration

During their brief stay in Kasanka National Park, Zambia, the millions of straw-coloured fruit bats, *Eidolon helvum*, disperse in all directions for their nightly forage.

Little red flying foxes, *Pterpopus scapulatus*, here in a small fig tree by an Australian river, are perhaps nomadic rather than truly migratory.

patterns. Again, satellite tracking has allowed individual bats to be followed as they travel hundreds of miles between roosts, taking them across country borders and even seas.

In Australia several of the larger fruit bat species migrate. Here, until recently the grey-headed flying fox, *Pteropus poliocephalus*, was only a seasonal visitor to the Melbourne area of Victoria. Now many are resident, but still the species carries out major movements throughout its range in eastern Australia as it follows the changing production of its favoured fruits and flowers. This species can utilize more than 100 roosts and travel over 3,000 km (1,900 miles) in a year. Since these movements are not strictly between A and B, but much more opportunistic, they may be regarded as more nomadic than strictly migratory. The grey-headed flying fox often roosts with the little red flying fox, which has long been regarded as nomadic since it is always on the move, although it doesn't cover the same geographical range as its larger relative. But this species has been shown to use up to *c*.40 different roosts in a year and to cover about 6,000 km (3,800 miles) between them. The third 'migratory' species in Australia, the black flying fox, *Pteropus alecto*, does not associate with the other species quite so readily but may still cover up to 2,000 km (1,300 miles) in a year between about 40 roost sites.

Other tropical fruit bats (and insectivorous bats) move seasonally between habitats that offer better feeding opportunities. In some cases, it may be that there is not more or better food available elsewhere, but that for the period that fruit and flowers are present in an area some species will move in and take advantage of less competition for their availability.

Quite unrelated to the Old World fruit bats are the fruit and flower feeders of the Central and South American family of spear-nosed bats (Phyllostomidae). Three of these migrate from Mexico into the southern states of the USA. The Mexican long-tongued bat, *Choeronycteris mexicana*, migrates during pregnancy to Arizona and New Mexico to give birth in June/July, although some populations in New Mexico may be more-or-less permanent residents (some may overwinter there). The species may also undergo occasional 'irruptive' behaviour, such as in September 1946 when a major invasion of individuals and small groups appeared in many scattered localities around San Diego, California. Females of the more northern populations of the greater long-nosed bat, *Leptonycteris nivalis*, also migrate northwards towards the USA in June to August, but the extent of the migration varies from year to year, such that the Texas population can vary between none and 14,000 individuals, probably depending on the variation in flowering succession from year to year. In this species the young are born in Mexico in April to June and then travel north with their mothers; the males remain in the south of the range. The southern long-nosed bat, *Leptonycteris curasoae*, also migrates over 1,500 km (1,000 miles) from Mexico to the USA. There may actually be two reproductive populations in Mexico, one with a spring birth period which migrates south, and one with a winter birth period which migrates north. But these populations are not separable genetically. The northerly migrants demonstrate two routes for separate populations: one along the coast feeding on columnar cacti and the other along the foothills of the Sierra Madre Occidental feeding on paniculate agave. In line with flowering times, these inland bats move later than the coastal bats. Colonies of 12,000–15,000 may gather in US caves in mid-May and disappear by September; various studies, including DNA, and the wide range of stages of pregnancy on arrival, suggest that these bats may accumulate from a wide area. Similar long-range migration may occur in this species in Venezuela.

With altitude

Altitudinal migration is also recorded in a few bat species. Perhaps the best studied is the Hawaiian hoary bat, *Lasiurus cinereus semotus*, a distinct subspecies of the North American hoary bat. On Hawaii the hoary bat is most active during the reproductive season of May to September below 1,000 m (3,200 ft) elevation, whilst activity is generally much greater at higher elevations (to 3,600 m (11,800 ft) during the winter. The diet usually consists largely of moths, but there is evidence that at the higher altitudes the bats show a greater preference for beetles. Working at these higher altitudes, Frank Bonoccorso of the Hawai'i Volcanoes National Park has also found that the bats frequently flew into caves where they were probably feeding on a *Peridroma* moth species that shelters in the caves and in rock rubble in large numbers, an unusual practice for bats of this genus and one that, with their relatively long narrow wings, they are not well designed for. In the case of the Hawaiian hoary bat it may be that the bats are not moving roost site up to the higher elevations, but just making the journey for improved foraging on a daily basis; in which case perhaps it cannot be regarded as an example of a true migration. A related species the southern red bat, *Lasiurus blossevillii*, has a distinct subspecies on the Galapagos Islands, *L. b. brachyotis*, where this too performs seasonal altitudinal movements. On the mainland of South America, this species is also considered to be a migrant in the southern parts of its range in Argentina and Uruguay.

How

In homing experiments bats have been shown to use the Earth's magnetic field for orientation, and they may use it for longer navigation. The physical adaptations to migration are not particularly marked in bats. There is a tendency for many of the longer-range migrant species to have longer and narrower wings (a high aspect ratio), but there are plenty of exceptions. Since they tend to migrate in short hops (generally less than *c.*50 km, or 30 miles, a day) with daily or longer stop-overs for feeding along the way, there is little pressure to be very highly adapted. One of the long-range migrants in Europe, Nathusius's pipistrelle, *Pipistrellus nathusii*, a small 6–10 g (0.2–0.4 oz) species with quite broad wings for a pipistrelle bat, has recently come under increased study with some interesting results. A juvenile male ringed at a migration point in Latvia in

Nathusius's pipistrelle, *Pipistrellus nathusii*, one of the few noted long-range migrants in Europe.

August 2015 was re-trapped in Southeast England after a journey of 1,453 km (903 miles) west–southwest in 50 days, an overall average of 29 km (18 miles) per day. Another male, also ringed in Latvia in August 2015, was recovered in Navarra, northern Spain, in March 2017, a distance of 2,224 km (1,382 miles) south–southwest and is believed to be the longest recorded movement of a ringed bat. More recently, an adult female Nathusius's pipistrelle tagged in Suffolk, UK, in spring 2021 flew directly to The Netherlands across the North Sea (over 200 km, or 125 miles) at a rate of c.50 km (31 miles) per hour (for which it must have had a favourable wind).

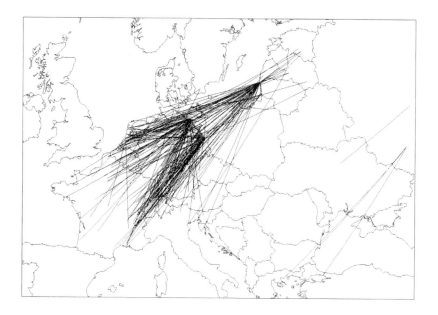

This 2005 map of the results of ringing of Nathiusius's pipistrelle, *Pipistrellus nathusii,* can be updated with recent records of movements between the Baltic states and Spain and between UK and the Baltic states and Russia.

CHAPTER 7

Predators

For the most part, bats are considered to be relatively predator-free. Inside the roost most bats are safe, although there are records of hibernating bats being taken by rats, martens (*Martes* spp.) and in at least one situation in Hungary repeatedly by great tits, *Parus major* – a quite small bird, but one noted several times as demonstrating carnivory, even cannibalism. There is more danger as the bats leave the roost, often through confined exits from caves or buildings. Here snakes, and even frogs, may be waiting at cave mouths, and cats, magpies (*Pica pica*), owls or other birds of prey may learn to pick the bats off as they sample the evening air before setting off on an evening's hunt.

Specialized predators

As the five million wrinkle-lipped bats, *Chaerephon plicatus,* emerge from Deer Cave in Borneo, large groups form a tight ring, a 'doughnut' of bats, that moves up the cliff face, making it extremely difficult for any would-be predator to pick one off. But then the ring breaks into a line that joins the general stream of bats moving away from the cave. Sometimes, no bat will take the lead and break the ring, and the ring of bats drifts slowly off towards the foraging grounds. The main great stream of bats will often be attended by a few birds of prey, which may take their fill but leave plenty more bats. One or two of the birds of prey are quite specialized bat predators and hence called 'bat hawks', notably *Macheiramphus alcinus* in Africa and Southeast Asia. There is a long list of birds other than birds of prey that have been recorded taking bats, but these are all rather adventitious occurrences. Nevertheless, there are a number of reports of an owl having learnt

The greater naked bat, *Cheiromeles torquatus*, of Mexico is parasitized by an earwig and a flea.

how to pick off the emerging bats, and repeatedly returning to a particular colony and doing irreparable damage to the colony. Predation on foliage-roosting bats is also quite regular; the introduced brown tree snake, *Boiga irregularis*, on the island of Guam is thought to have been a major contributor to the extinction of the endemic flying fox, *Pteropus tokudae*, by eating young bats, for example. Another predator of the large flying foxes is crocodiles. For many of the larger fruit bats that form colonies in trees, the first thing they do on leaving the roost in the evening is to drink. They may take water directly into the mouth but usually just do a 'touch and go' and dunk their chest fur into the water. Most of the time they are safe, but in some Australian rivers freshwater crocodiles are waiting and try to grab the bats with their jaws as they fly past. And we should note here that some of the larger carnivorous bats will also sometimes include smaller bats in their diet.

A ring of wrinkle-lipped bats, *Chaerephon plicatus*, circles up the cliff face before joining the main stream of bats emerging from their cave entrance.

An Australian saltwater crocodile taking a little red flying fox, *Pterpopus scapulatus*, as it flew out of its roost along the Daintree river mangroves, northern Queensland.

In itself most of these acts of predation have little impact on bat populations, but taken together they may make a significant contribution to the annual mortality. There is, of course, a trade-off here – you can emerge relatively early and risk exposure to predation, or wait until darkness and miss the best feeding opportunities during the dusk peak of insect activity. That it is worth avoiding predation as far as possible is demonstrated in areas where there are no predators. The Leisler's bat, *Nyctalus leisleri*, of Europe is generally nocturnal, but its very close relative on the Azores in the mid-Atlantic, *Nyctalus azoreum*, spends a lot of time foraging for insects during the day. At the other end of the world, the common Samoan flying fox, *Pteropus samoensis*, is largely diurnal on Samoa where there are no predators, but it is a lot less diurnal on Fiji, where the peregrine falcon also exists.

In many parts of the world, particularly in Europe, domestic cats are a major predator on bats. And there is, of course, one major predator that will be dealt with later – *Homo sapiens*.

Parasites

Bats carry an extraordinary array of specialist invertebrate ectoparasites. This may partly be a result of their colonial nature, but the parasites also have to develop strategies that: allow for their hosts to be present in the roost for only part of the year; cater with the absence of a nest where they can remain close to the host and where their larvae can develop; , that enable them to hang onto a bat in flight and, perhaps most importantly, that enable them to defend themselves from the assiduous grooming of their hosts.

For the most part the animals referred to in this chapter are indeed parasitic, as they live at the expense of the host. In a few cases it is difficult to decide whether they are using the host for all of their food or just for transport (in which case they are described as phoretic) or to be close to by-products from the host that it can use as food (in which case they are described as commensal). In a few other cases the host may actually receive some benefit from the presence of its guest, and then the relationship is described as symbiotic. But for the purpose of this chapter we can call them all ectoparasites.

Despite the diversity of insect ectoparasites, there is one glaring omission: there are no lice on bats, and lice form by far the largest group of insect ectoparasites and are extremely widely distributed on birds and other mammals. Instead, there are fleas (all in one family, Ischnopsyllidae), two families of bat flies (and two other families of flies, each with a single species, whose relationship with the bats is less parasitic), two families of true bugs (one exclusively on bats, and one with most species on bats, a small number on birds and one species being the human bedbug, *Cimex lectularius*), and one family of earwigs. There is also a very wide array of mites associated with bats for at least part of their life cycle. Most of these parasites are highly specialized, either in their structure or in their life cycle (or both), for their life on a bat.

These groups of parasites show a wide range of life cycles. The fleas have a complete life cycle of egg–larva–pupa–adult. The larvae are free-living, scavenging around the roost and the guano below their hosts. When the adult fleas emerge on the floor of the roost they must find their way onto a bat, which may involve

A wingless bat fly (*Penicillidia* spp.) on the head of a long-winged bat, *Miniopterus mossambicus*, Gorongosa National Park, Mozambique.

crawling up to the colony, but often that is just not practical and so they hitch a ride on a youngster that has dropped from the colony or from its mother when she comes to collect it. Once on the host they are well designed to remain there. For the most part, the bug family that includes the human bedbug (the Cimicidae) use the hosts more as a restaurant; they spend most of their life off the hosts, hiding around the roost and only venture onto the hosts to feed. They lay eggs off the host and they go through several immature nymphal stages before becoming adult. For this life cycle they do not need the great modification that is required of parasites that travel with the host. Members of the related family Polyctenidae, however, live all their life on the host – their eggs hatch internally and the female gives birth to a series of early nymphal instars. The bat fly family Nycteribiidae, although quite unrelated, has a somewhat similar cycle, in that the egg hatches internally and the larva goes through the whole of its development inside the female abdomen. However, the bat fly then moves off the bat and deposits its larva or pre-pupa somewhere nearby around the roost, making sure it is well fastened to the substrate. Here the larva immediately pupates and continues its development inside a shiny hemispherical pupal case. Most of the related family of bat flies, the Streblidae, do much the same. The earwigs, like the bedbugs, have an incomplete metamorphosis and so go through a number of nymphal stages before becoming adult. As nymphs they are more likely to remain in the guano below the roost, but they are equally happy to gain access to a bat.

One of the most important tests for most of these parasites is to stay on the host (although they may want to travel between hosts). For this they have a number of special features common between them, and since the groups of parasites are often quite unrelated, they provide good examples of convergent evolution: mostly they are flightless; they are usually flattened laterally or dorsoventrally to allow them to slide through the fur of their host; they have very robust claws; and most have elaborate combs of spines and other structures to protect their body joints and to hang onto the host's hairs and help stop them being dislodged. The nycteribiid bat flies are barely recognizable as flies; they are wingless and the thorax is greatly distorted and flattened, so that they look like a rather odd spider scurrying amongst the bat's fur. The bat fly family Streblidae displays an extraordinary range of forms. Many look like fairly typical flies, but in the New World there are genera that have lost their wings completely or almost completely, one genus that rolls its wings up into a tube and tucks them under some long stiff bristles to keep them out of the way, and others that carry a fascinating array of combs of spines and other features to stop them being dislodged from their host. But perhaps the most bizarre streblid genus is in the Old World tropics. *Ascodipteron* emerges as a fairly normal-looking fly, but the female develops a large cone-like head that enables it to push under the skin of the bat, usually on the forearm or around the head. As it burrows under the skin it sheds it wings and legs, the abdomen enlarges into a sack that engulfs the head and thorax and just leaves the posterior end of the fly projecting from the skin of the bat and from which will be jettisoned the pre-pupa as they are developed. The earwigs seem the least well-adapted to travel on a bat, especially since they live on the naked bat, *Cheiromeles*, but they hang on remarkably tightly, particularly in the folds of skin, and must have some very special structures on their feet to enable that.

The earwig, *Arixenia esau*, lives in the debris below the bat roost, but will equally spend time on its host, the greater naked bat, *Cheiromeles torquatus*, of Southeast Asia.

The mites (including ticks) also offer a bewildering array of forms and life cycles, some with only a part of their life cycle as a parasite. Thus, only the larvae (the first instar) of trombiculids live as parasites; otherwise they are free-living. Other mites visit the bats at each stage of their development, and yet others are firmly committed to the bats for their entire lifetime. There are mites that specialize in living in the fur, others on the wing membrane, some inside the nostrils, or sliding up and down the large sensory hairs on the face of a horseshoe bat, or living in little sacs on the wing membrane or along the forearm.

As the saying goes 'big fleas have little fleas upon their backs to bite them…', and it is true that fleas sometimes have mites living on them either as parasites or for hitching a ride, and nycteribiid bat flies can be attacked by parasitic fungi,

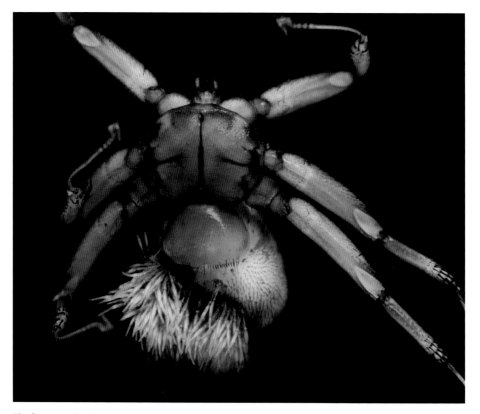

The fungus *Arthrorhynchus nycteribiae* is parasitic on the abdomen of the bat fly, *Penicillidia conspicua*, from a Daubenton's bat, *Myotis daubentonii*, Romania.

such as *Arthrorhynchus*. Not to be outdone, the bat flea *Lagaropsylla* attaches to a parasitic earwig, with anything up to 40 fleas on one earwig, to hitch a lift to the naked bats, which they share as a host.

All the insect ectoparasites described above are blood-feeders, except the earwigs. And they can occur in some profusion. It might be expected that all these parasites have a debilitating effect on their hosts, but on the whole they are tolerated. In the temperate regions of northern Europe one might find up to 15 bat flies on an individual bat. However, compared to many bats in the tropics, this is a small load; in the Neotropics it is possible to find several genera and quite large numbers of individuals (and a range of mites) on a single bat. However, it is true that sometimes high infestations may be a symptom of another problem and may affect particularly juvenile bats that are inexperienced at grooming them off.

There is also a wealth of endoparasites (parasites living inside the host, in its blood, tissues or alimentary canal), some of which can have a deleterious impact on the host and which transmit or cause disease. These include parasitic worms, bacteria, other protozoa, viruses and fungi.

Pathogens

WHITE-NOSE SYNDROME

In the winter of 2006 a number of dead and dying bats were found in caves in New York State, USA. It was quickly established that bats were awakening from hibernation and flying a lot around the entrance to the hibernacula, and they were developing a white fluff on bare skin areas, particularly on the nose, ears and wings. Many were severely emaciated and were at their worst in late winter, March and April. At this time, physiological functions, including immune responses, are at their weakest. The problem was dubbed white-nose syndrome (WNS), and the syndrome consists of fungal growth and behavioural changes combined with mass mortality. The white fluff was identified as a new fungus, originally described as *Geomyces destructans*, later transferred to the genus *Pseudogymnoascus*. Initially, despite detailed post-mortems, it was not known whether the fungus (now widely known as '*Pd*') was the cause of the disease, but eventually it was established that this cold-dwelling fungus was indeed the agent of the mass mortality through the fungus invading the subdermis and aggravating the bats, causing the arousals that result in starvation. Over the following years the syndrome was recorded as spreading further south and west, and by 2020 it had been recorded in 32 US

The white fungus on the nose and ears of these hibernating bats is characteristic of white-nose syndrome of North American bats.

states and seven Canadian provinces. And it has killed millions of bats. The bats affected are mainly *Myotis* species, particularly the little brown bat (*M. lucifugus*), but also endangered species such as the grey bat (*M. grisescens*), and the Indiana bat (*M. sodalis*). There was concern about where the fungus had come from and whether bats in other countries were at risk. Over the years it was established that the same fungus is widespread in caves and tunnels in Europe and is also found in Russia and China, but that the associated syndrome, including mass mortality, was not occurring; there the condition is referred to as white-nose disease. It is now clear that there is a baseline immunity in European bats, which have probably been living with the fungus for a very long time, whereas the North American bats did not have this immunity when the fungus was somehow introduced quite recently, probably by a human agency of a caver, cave tourist or bat worker. Strong collaborative work between researchers in North America, Europe and elsewhere is trying to establish the source and timing of the infection and to predict the future progress of the fungus in North America, and, of course, to mitigate its effects.

LLOVIU VIRUS

In June 2002, mass mortality of Schreibers' bent-winged bat, *Miniopterus schreibersii*, was found in caves in Spain, Portugal and southern France. Large numbers of corpses were found (up to 1,400 in one cave in Spain) and whole colonies were missing.

Specimens were collected for post-mortem, but the specimens were not fresh and so a full post-mortem was not possible. Some of these corpses and internal organs were stored. It was noted that previous unexplained mass mortalities had been recorded in this species from Italy and what is now regarded as a closely related but separate species in Australia. In the following years many bats of this species, and another found in the same roosts, were sampled with oral and anal swabs. A virus was isolated from a few of the *M. schreibersii* swabs and retrospectively from the tissue samples retained from the 2002 incident. Using molecular means this was shown to be a new filovirus (the group of viruses that includes Ebola) and the first such virus in Europe. It was named Lloviu (LLOV) after the cave in which it was first collected. It has not been possible to show that this was the causative agent of the mass mortality, but the virus has subsequently been found in Hungary, where it was also associated with significant mass mortality events of *M. schreibersii* in 2013, 2016 and 2017 (the latter two quite minor events). Again, it cannot be shown that the virus was the causative agent, but the Hungarian virus is a very close variant of the Spanish LLOV.

There have been other mass bat mortalities, mostly unexplained and particularly in Pacific island flying foxes, and a growing list of viruses has been identified from bats, most of which would appear to be of no animal (or human) health importance. Generally, the impacts of viruses and other microbiological agents and endoparasites on bats is not well known, but for the most part the host and parasites must exist in relative harmony, apart from when some balance goes wrong, which results in events like mass die-offs.

ZOONOSES

Traditionally, most of the studies of viruses and other microbiological agents of wild animals were directed at rodents and ungulates because they were more accessible and because of their role or potential role in the spread of disease to humans or domestic animals. The recent interest in bat viruses has been encouraged by the growth of molecular and other techniques for studying the viruses and the increased ability to sample bats in the field. Coronaviruses were hardly recorded from bats until the emergence of the SARS-CoV coronavirus in 2002, but now people have started to search for them; they prove to be very common and in most cases very specific in their host. Where these viruses impact on human health, the associated diseases are referred to as zoonoses; we will look a bit more at some of these viruses later (see p.131).

CHAPTER 8

Classification

The classification of bats has been through some major upheavals in recent years, particularly as a result of the use of DNA to reassess the relationships between groups and between species. As a group, bats have a long history, a rich diversity and the ability to occupy a wide range of geographical areas and habitats.

Systematics, evolution and history

The earliest fossil bats come from the Eocene, more than 50 million years ago. The oldest is *Onychonycteris finneyi* from Wyoming at *c.*52 million years ago, which although clearly capable of true flight does not display any features associated with the use of echolocation. So maybe it answers an age-old question as to which came first, flight or echolocation – flight. *Onychonycteris* would also appear to have been a good climber and so may have flown to places where it could scramble around to find food, much as a few bats do today. But whether it was nocturnal or diurnal remains a mystery. There are a number of other very well-preserved fossils from only a little later (e.g. *Icaronycteris*) that are also clearly flying bats with well-developed auditory systems that would suggest that some bats were echolocating by that time. *Icaronycteris* is based on the species *I. index* from a few localities in North America, but a further species has been described from India and a possible species of this genus is described from France. Thus it appears that while *Onychonycteris* did not use echolocation, *Icaronycteris* and other fossil bats of this era are, in general, much like our current insectivorous bats.

There is actually quite a rich fossil record for bats, including some really well-preserved early specimens, although many of the taxa described are based

Icaronycteris index from Wyoming, USA, one of the first complete early Eocene fossil bats to be described, used echolocation to catch prey, probably mainly by perch-hunting.

Palaeochiropteryx tupaiodon is one of many well-preserved fossils of more than 45 million years ago from shale pits at Messel, Germany. Even stomach contents can be examined.

on isolated jaws or teeth. The range of fossil material naturally paints a very incomplete picture of their evolution and tells us very little about the origin of bats but does show an early development for some of our currently recognized families, some families that have disappeared and some groups showing a very different distribution from their current situation. There may have been major periods of diversification during the Early Eocene Climatic Optimum and that of the Mid-Miocene, and the following are a few examples of what is known of the development of particular bat families.

The rather poor fossil record of the Old World fruit bats (Pteropodidae) begins with one described from the Eocene in Thailand, but the fossil is only one tooth and there has long been controversy about whether it really is from a fruit bat. *Archaeopteropus transiens* was described from the Oligocene (*c.*32 million years ago) in Italy, and although it was a relatively complete fossil (the original specimen is lost, but some casts remain), there has been some suggestion that it, too, may not be a true pteropodid. The Miocene period (<23–5 million years ago) covers a long period and there are fossil pteropodids from Africa, Asia and Europe (*c.*12 million years ago) from this period. The findings in Europe support the idea of a western movement of pteropodids from Australasia and Asia, through Europe, to give rise to the African fauna.

Fossil horseshoe bats (Rhinolophidae) are recorded from the Miocene, Pliocene and Pleistocene. An older genus, *Vaylatsia* from the Late Eocene and Oligocene in Europe, is now regarded as from the closely related family of Old World leaf-nosed bats (Hipposideridae). The key fossils from Europe do not provide much information about the origins of the group, but there was likely to have been a rapid increase in diversity in the Miocene of *c.*18–15 million years ago, while a proposed origin in Asia is not supported by current molecular analysis, which points to an African origin.

For the plain-nosed or vesper bats (Vespertilionidae) there are a number of genera recognized from the Eocene and Early Oligocene, which belong to this or a closely related family, with the earliest definite vespertilionid, *Premonycteris vesper*, from the Early Eocene in France. Subsequently there are many fossil taxa described that suggest that the family may have had a European origin, with the first fossils for North America from the Late Oligocene, and for Africa and Asia from the Middle Miocene. The colonization of the Neotropics and Australasia would appear to have been quite recent. The largest genus of bats, *Myotis*, has over 120 extant species and over 40 extinct species. The late G. F. Gunnell and colleagues (especially Nancy Simmons of the American Museum of Natural History, New York) have done much of the pioneer and synthesis research on the fossil bat fauna.

The current diversity

The recently published *Handbook of the Mammals of the World, Vol 9: Bats* recognizes over 1,400 current bat species arranged in 21 families. As a measure of the increased interest in bats and our ability to study them in various ways, this is a very marked increase from the 950 species listed in *Mammal Species of the World* in 1993 and 1,150 in a revised version of the same work in 2005. Bats are then a very successful group in developing a diversity of species and behaviour.

They have also been very successful in colonizing the world – helped, of course, by their power of flight. While bats do live north of the Arctic Circle, they have not been able to establish in the Arctic proper, nor on Antarctica and its islands. Otherwise, they have colonized all major land masses and many islands, being absent from those in the South Atlantic on the Mid-Atlantic Ridge and occurring only as occasional visitors to Iceland. They occur on most of the islands of the Indian Ocean, although not the Chagos Archipelago in the centre, nor the islands of the southern Indian Ocean. In the Pacific Ocean they occur from South America to the Galapagos Islands in the east. From the west they occur as far as Hawaii in the North Pacific and to Fiji, Tonga, Samoa and the Cook Islands in the South Pacific. A very high proportion of the remote island species are endemic, and many are Old World fruit bats, especially of the genus *Pteropus*.

For a very long time the bats were split into two main groups or suborders: the Old World fruit bats (Pteropodidae), called the Megachiroptera, and the rest, called the Microchiroptera. This classification was barely questioned until towards the end of the 1980s, when a group of researchers led by Jack Pettigrew suggested that these two groups were not monophyletic (i.e. that they did not have a common ancestry) and that the Megachiroptera were more closely related to primates than to the other bats. This was on the basis of a number of characteristics including such features as the nature of neural pathways. This formed the topic of a very 'robust' debate (a 'full and frank discussion') at the 8th International Bat Research Conference in Sydney in 1989, but with no resolution except to leave things as they were. However, within ten years, with ever growing ability in molecular studies, this classification was again being questioned. And there is now growing acceptance that there are indeed two suborders, but comprising of one for the Pteropodidae with six other families (Rhinopomatidae, Craseonycteridae, Megadermatidae, Rhinonycteridae, Hipposideridae and Rhinolophidae) as the Yinpterochiroptera, and the other 14 families in the Yangochiroptera. Essentially, although the traditional groupings into superfamilies has become unsettled in recent years, the family delimitations have remained pretty stable over the years, but just recently three new small families have appeared, largely through molecular studies encouraging more detailed studies of the characteristics of these bats. These are the trident bats (Rhinonycteridae), the wing-gland bats (Cistugidae) and the bent-winged bats (Miniopteridae). Thus, we now recognize 21 families, and it is interesting to note that only three families (the Emballonuridae, Molossidae and Vespertilionidae) occur in both the Old and the New World; all the others, including some species-rich families such as Pteropodidae, Rhinolophidae and Phyllostomidae, occur only in the Old World or the New World. We will look at all the families of bats in more detail.

So, if we accept that bats are a monophyletic group of species, then they probably arose from a small arboreal insectivorous ancestral stock, probably in the Palaeocene or Mid- to Late Cretaceous period, which stretches back from 60–100 million years ago. And most of the families we recognize today were separated by the end of the Eocene.

Genetics

The study of the systematics and other features of the lives of bats has been greatly advanced in the last 20 or 30 years by the development of techniques in DNA and other molecular analysis. Studies of the greater horseshoe bat, *Rhinolophus ferrumequinum*, and its relatives can be used to demonstrate this.

It came as a great shock to many bat workers (and still is to some) that the family of horseshoe bats (Rhinolophidae) is more closely related to the fruit bats (Pteropodidae) than to most other bats. Molecular studies have demonstrated this relationship, and the relationship of five other families, grouped into the superfamily Rhinolophoidea, which together with the Pteropodidae form one of the two suborders of bats (Yinpterochiroptera). These studies, using the degrees

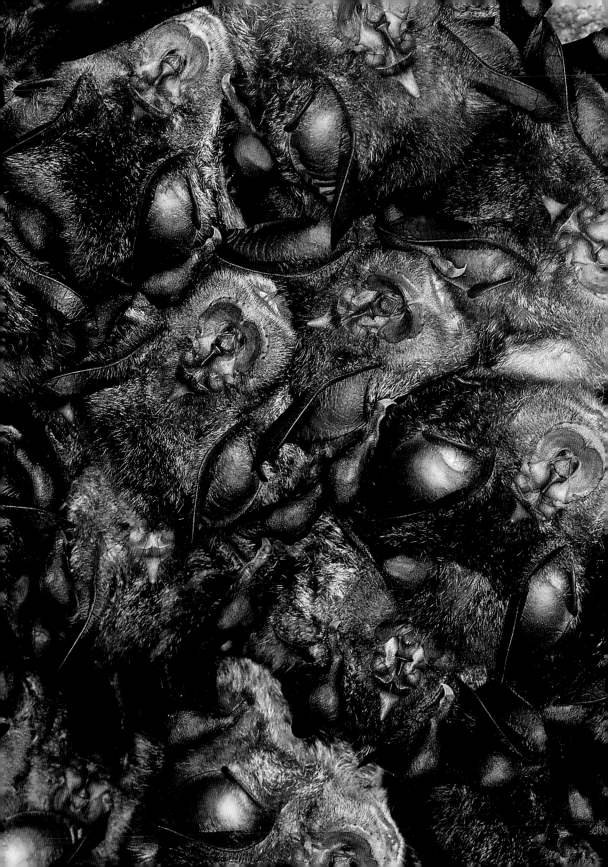

of difference, also tell us that the separation of the Pteropodidae came somewhat later than had been expected at *c.*59 million years ago and that the separation of the Rhinolophidae from its nearest relatives (the Old World leaf-nosed and trident bats) came *c.*42 million years ago.

Studies also support the idea of a single genus in the Rhinolophidae, with some divergence of opinion as to whether the genus originated in Asia or, more likely, Africa. Over the years the genus has been separated into groups of species considered to be closely related. Further to assisting with the delimitation of species, DNA studies also tell us about the levels of difference between species and thus have supported some of those species groups, but not all. They have also enabled the separation of a large number of extra species, such that while nearly 70 species were recognized 20 years ago, that number has now risen to 109.

DNA has been used to examine the way a species has spread geographically over time and how that distribution has been affected by, for instance, climate change, to plot where there were glacial refugia and the progress of recolonization from there.

At the species level molecular studies can give information about the relationships within and between colonies, about the levels of gene flow and to what extent that relies on the mating strategy. Thus the summer maternity colonies of greater horseshoe bat tend to comprise quite closely related females, but these meet with a number of males, which may or may not be related, in the autumn mating sites. It can also be shown that there are males that get more than their fair share of the mating ('megastuds') and there are females that produce more than their fair share of young ('supermums'); it can also be seen how much cheating occurs in species with a harem structure (such as the greater spear-nosed bat, *Phyllostomus hastatus*). In suitable colonies, analysis of DNA from droppings can provide further information on the number of individuals in the colony (including the level of turn-over in the colony) or the number of males mixed into the colony. Apart from the information about the bats themselves, DNA is increasingly being used to provide information on diet from analysis of droppings.

Studies of DNA and other molecular studies have revealed a wide range of information on the history and life-style of bats, like these greater horseshoe bats, *Rhinolophus ferrumequinum.*

Family accounts

The bats are currently divided into two suborders, five superfamilies and 21 families. A brief account of each of the families follows. [NB There is also a quick reference table of the families on p.154 using the information in the first line under each account here.]

SUBORDER: Yinpterochiroptera
SUPERFAMILY: Pteropodoidea
FAMILY: Pteropodidae. Old World Fruit bats. 46 genera, 191 species. Old World tropics. Fruit- and flower-feeders.

Includes the largest bat species, weighing up to 1.5 kg (3.3 lbs) and with a 1.7 m (5.6 ft) wingspan. Also includes some small species of *c.*12–20 g (0.4–0.7 oz) and a wingspan of 250 mm (9.8 in). The largest genus, *Pteropus*, includes 63 species, nearly all of which are endemic to particular islands or groups of islands. Mainly tropical from West Africa and its islands east to the Cook Islands in the Pacific, north to Ryukyu islands of Japan, Himalayas of Khatmandu in Nepal, and coastal Turkey and Cyprus in the Mediterranean. Mostly with rather dog-like heads, such that the larger species are generally known as flying foxes, which feed on fruit and flowers. Many of the smaller species have more elongate muzzle and tongue and are nectar specialists. Important pollinators and seed dispersers for a wide range of plants. Some species are notable migrants, following succession of fruiting and flowering. Mostly roost in foliage, often with major colonies (or 'camps') in canopy trees, some species roosting in caves and sometimes forming large colonies.

SUPERFAMILY: Rhinolophoidea
FAMILY: Rhinopomatidae. Mouse-tailed bats. 1 genus, 6 species. Old World tropics and northern subtropics. Insectivorous.

Arid zones of northern Africa north to the Mediterranean in the west and across the southern Palaearctic region to South Asia, where they occur through most of India. Records from east of here need confirmation. Small to medium-sized bats of 5–40 g (0.2–1.4 oz). Very obvious long, free, mouse-like tail, ears joined above the head and characteristic nosepad. Insectivorous, at times dominated by small to medium-sized beetles taken by aerial hawking. Mostly small colonies of 20 or less, some colonies to 20,000. Smaller colonies usually in rock fissures, larger colonies in caves, tunnels, tombs and other artificial structures. Seasonal movements with winter torpor in some areas.

FAMILY: Craseonycteridae.
Hog-nosed bats. 1 genus, 1 species. Thailand and Myanmar. Insectivorous.

Considered most closely related to mouse-tailed bats and false vampire bats. Only known from a small area of western Thailand and southeastern Myanmar. Some suggestion that the Myanmar population may be a second species has not been confirmed. Arguably the smallest bat in the world (and possibly the smallest mammal) weighing 2–3.2 g (0.07–0.11 oz). It has a distinct bulbous nosepad, somewhat hog-like in appearance. Insectivorous, feeding on a variety of prey, but especially small moths and beetles. Roosts in limestone caves, where colonies usually small, about 10, with up to 100 currently known and historic records of up to 500. Colonies not clustered, with bats hanging individually.

FAMILY: Megadermatidae.
False vampire bats. 6 genera, 6 species. Old World tropics to northern Australia. Insectivorous, some partial carnivory.

A small family of species scattered from Africa (two species) through Asia (three species) to northern Australia (one species). Medium-sized to large bats weighing 20–170 g (0.7–6 oz), with the African species at the lower end of the size range and the Australian ghost bat being one of the largest bats outside the Pteropodidae. They have very large, rounded ears that are joined together at the base, and a large relatively simple noseleaf that covers the muzzle. They also have very large broad flight membranes, but no external tail. Mainly feed on insects and other invertebrates, but at least three species (heart-nosed, great Asian false vampire and ghost bats) all take a significant amount of vertebrate food including frogs, lizards, fish, birds or small mammals, including other bats. Prey is usually taken by gleaning from vegetation or the ground or by forays from a feeding perch. Roosts vary from free-hanging in vegetation (yellow bat), or in tree hollows or shallow caves (heart-nosed bat) in African to caves and temples (occasionally hollow trees) in Asia, and caves and mines in Australia. The yellow bat roosts singly or in pairs, most others in smaller colonies of up to 100, but the ghost bats and great Asian false vampire are recorded to sometimes form larger colonies of up to 1,500 or even over 2,000, respectively.

FAMILY: Rhinonycteridae.
Trident bats. 4 genera, 9 species. Gulf of Oman south through (mainly East) Africa to southern Africa, northwest Australia.

A recently recognized family, long considered a discrete group on the structure of the noseleaf, but only recently separated from the family Hipposideridae on the basis of modern phylogeny, supported by molecular genetics. Mainly tropical from either side of Gulf of Oman and Yemen through Eastern Africa (with an outlier in west coast Africa) through most of southern Africa, also Madagascar and Aldabra Atoll in the Indian Ocean. One species in northern (mainly northwest Australia). Small bats weighing 5–20 g (0.17–0.7 oz). All the Afrotropical species have a distinct trifid process arising from the posterior leaf of the nose-leaf. Percival's trident bat (*Cloeotis percivali*) has possibly the highest frequency echolocation calls of all bats (at over 200 kHz). Insectivorous, mainly feeding on small to medium-sized moths, but taking other prey, such as beetles and termites opportunistically. Roost in caves or other underground sites, often in large numbers.

FAMILY: Hipposideridae.
Old World leaf-nosed bats. 7 genera, 88 species.
Old World tropics and subtropics. Insectivorous.

Closely related to the horseshoe bats, but with a somewhat different, and sometimes more complex noseleaf. Occurs in the Old World tropics and subtropics north to North Africa and across to southern China and the Ryukyu Islands of Japan, south to much of southern Africa and the northern half of Australia. Wide size range from 5 to 150 g (0.17–5.3 oz). Insectivorous, feeding on a wide range of insects partly related to the size of the bat species, and with the possibility that one or two of the larger species might take small vertebrates such as frogs. Can become torpid, or even hibernate, at the northern ends of the range. Mostly roost in caves in small to large numbers, usually hanging slightly apart; a number of species have moved into buildings and other artificial structures.

FAMILY: Rhinolophidae.
Horseshoe bats. 1 genus, 109 species. Old World north to *c*.55° in Europe and 45° in Japan and south throughout Africa and the eastern edge of Australia. Insectivorous.

Closely related to the Old World leaf-nosed bats (and trident bats), but with a distinct horseshoe-shaped flap on the noseleaf. Mainly Old World tropics and subtropics but with species reaching into temperate zones to Wales (UK) in the West and northern China and most of Japan in the east. In the south reaching the bottom of South Africa and southern Australia in Victoria. Small to medium sized, from less than 5 to *c*.50 g (0.17 to *c*.1.7 oz). Insectivorous, feeding on a wide range of insects partly related to size of bat species. Hibernate at extremes of latitude. Traditionally mostly cave bats, but some species have adopted buildings or other artificial structures. A few species roost in foliage. Colonies can be small, especially in foliage-dwelling species, but can be tens of thousands in caves and similar underground structures

SUBORDER: Yangochiroptera
SUPERFAMILY: Emballonuroidea
FAMILY: Emballonuridae.
Sheath-tailed bats. 14 genera, 54 species. Through
Old and New World tropics and subtopics
north to Turkey and south to Victoria, Australia.
Insectivorous.

Most closely related to the Nycteridae, but quite
different looking bats. One of the three bat families that occur in both the New
World and the Old World. Mainly tropical but, in the Old World, spreading
through the subtropics north to Turkey in the west and China in the east, south
to the cape coast of South Africa and Victoria, Australia. In the New World,
from Mexico south to southern Brazil. Also many islands of the southwestern
Pacific. Small to medium-sized/large bats, weighing 3–100 g (0.1–3.5 oz), with
a plain nose (no facial adornments) and when at rest with the distal half of the
tail protruding from the tail membrane (in flight an elastic sheath allows the
tail to be drawn back into the tail membrane). Wing sacs are present, mainly in
males, of some New World species, and other glandular sacs on throat, chest or
along tail are present in some other species. Insectivorous; despite a few species
being quite large there is no evidence of carnivory. Wide range of roost sites
with some species roosting on the trunks or branches of trees or in tree hollows,
behind palm leaves, in rock crevices and in caves, tunnels, tombs and other semi-
underground artificial sites. Colony size is usually small, from just a few bats,
particularly in those species roosting on trees, to about 50, or to many thousands
in a few of the cave dwellers.

FAMILY: Nycteridae.
Slit-faced bats. 1 genus, 15 species. Africa, two species
Southeast Asia. Insectivorous, partly carnivorous.

A small and quite distinctive and uniform group of bats,
looking unlike their nearest relatives, the sheath-tailed
bats. Most species live on the African mainland, with
one species extending up into the Middle East, most
species with quite a restricted range; one hardly known

species on Madagascar and two species in Southeast Asia. Small to medium-sized bats, weighing 4–43 g (0.14–1.5 oz). Characterized by large ears and a long deep furrow along the top of the muzzle from the nostrils; this hollow has fleshy outgrowths alongside and partially covering it. The flight membranes are very broad with the tail ending in a T-shaped bone at the margin of the tail membrane. Insectivorous, including a range of other ground invertebrates such as scorpions and spiders, with the largest species taking a significant proportion of small vertebrate prey from fish, frogs, mammals (including other bats) and birds. Usually roost in shallow caves and rock hollows; hollow trees are also common, and they sometimes use artificial tunnels and cellars; one species regularly roosts in aardvark burrows. They may roost singly or in pairs, or, in most species, in small colonies of up to c.10 bats. Colonies of 50–100 are rarely recorded and up to 1,000 is known.

SUPERFAMILY: Noctilionoidea
FAMILY: Myzopodidae.
Madagascar sucker-footed bats. 1 genus, 2 species.
Madagascar. Insectivorous.

Restricted to Madagascar, with one species in the drier west and one in the wetter east. Small bats weighing 7.5–10.5 g (0.26–0.37 oz). Characterized by well-developed adhesive pads on the wrists and ankles. The ears are large and broad, and the upper lip is wide and wrinkled. Insectivorous, feeding mainly on moths and cockroaches, with a range of other insects and spiders also recorded. They use the pads on the wrist and ankle as a wet adhesion system to roost by hanging onto the smooth inner surface of the furled leaves of palms such as *Ravenala* and *Bismarckia*, which makes them change roost site frequently as the leaves unfurl. They roost head-up, which may be a consequence of the way the pads work. Usually found roosting in very small numbers, often singly, but with groups of up to 32 recorded.

FAMILY: Mystacinidae.
Short-tailed bats. 1 genus, 2 species. New Zealand.
Insectivorous, with some fruits and flowers.

The relationships of this small family are uncertain, but it is currently considered to belong to the Noctilionoidea. One species is widespread mainly New Zealand (both the North and South Islands. The other species, described from the islands off the south coast of South Island, has not been seen since the late 1960s. Medium-sized bats of 10–25 g (0.35–0.88 oz). An enigmatic bat, which superficially is not very distinctive externally but does have a number of particular features, mainly associated with its terrestrial behaviour. The muzzle is long and extends beyond the lower lip; the nostrils are rather prominent and tubular on a rudimentary narial pad and surrounded by stiff bristles. The wings can be folded with the tips tucked into folds along the side of the body by the thigh, and the hindlimbs are very robust with large claws that have a small subsidiary denticle or talon under the base (also found on the strong thumb). The tail protrudes partway through the tail membrane, and the distal area of the tail membrane can be rolled forward to lie against the body. Mainly insectivorous and taking a wide range of aerial or ground-dwelling invertebrates, but also taking fruit and nectar and an important pollinator of the ground flowering wood rose (*Dactylanthus*). Has been recorded from caves, but now seems to roost almost exclusively in tree holes. Isolated bats may roost in a range of other situations. Colonies usually of several hundred (and up to more than 6,000 recorded), but smaller colonies also occur and the larger colonies may disperse to roost more individually during winter.

FAMILY: Noctilionidae.
Fisherman or bulldog bats. 1 genus, 2 species.
Central and South America. Insectivorous and carnivorous (fish).

Central and South America from northern Mexico to Paraguay and northern Argentina; also on many

Caribbean islands. The two species have a similar distribution except that the smaller species is absent from most of Mexico and the Caribbean. Medium-sized to large species, with the lesser bulldog bat weighing 20–45 g (0.7–1.6 oz) and the greater bulldog bat weighing 50–90 g (1.7–3.2 oz). A largely bare face with heavy jowls and a split upper lip, or 'cleft lip'. Ears long, narrow and pointed. Long legs with huge feet and huge claws. A well-developed tail membrane is supported by large calcars (spurs of cartilage) and the tail ends where it protrudes slightly at about half the length of the tail membrane. The fur is extremely short and velvety, bright orange–yellow in male, duller yellowish-brown in female. Both species are insectivorous, but the greater bulldog bat is largely piscivorous (fish-eating) and the lesser bulldog bat rarely takes fish. The invertebrates taken are varied and include scorpions, shrimps and small crabs (greater bulldog bat), and the lesser bulldog bat also eats some fruit. Roosts generally in trees, but caves and other artificial semi-underground structures are sometimes used. Smaller groups may comprise a male and a group of females, or may be a group of bachelor males. Otherwise larger colonies of *c*.30 to several hundred may form.

FAMILY: Furipteridae.
Smoky bat and thumbless bat. 2 genera, 2 species.
Central America to southern Brazil, N Chile.
Insectivorous.

The smoky bat (*Amorphochilus*) is found only along a narrow strip of the Pacific coast from Ecuador to northern Chile; the thumbless bat (*Furiptera*) only from southern Nicaragua through Central America and much of the territory east of the Andes south to northern Bolivia and southern Brazil. Small bats: 3–10 g (0.1–0.35 oz). Both species are more-or-less thumbless, with the rudimentary thumb incorporated into the flight membrane and no claw. The muzzle is blunt and well furred with an upturned nose; the ears are large and funnel-shaped with the base partially covering the tiny eyes; the single pair of mammary glands are located low down the body close to the genitalia; the legs are long with a very long tail membrane that extends well beyond the end of the tail. Insectivorous, mainly taking moths, but also beetles and flies. Roost in caves, where they often prefer small cavities or crevices, but are also found in hollow logs and a range of artificial structures (buildings, tunnels, cellars). Colonies are often small groups of five to ten individuals and can be up to 250 (thumbless bat) or 300 (smoky bat), but individuals stay slightly apart from others and have no contact.

FAMILY: Thyropteridae.
Disc-winged bats. 1 genus, 5 species. Central and
South America south to central/southern Brazil.
Insectivorous.

From Central America (southern Mexico, Belize) south to East and Southeast Brazil (Rio de Janeiro). Small bats weighing 3.5–6 g (0.12–0.21 oz). Characterized by very distinct suction pads on the soles of the feet and the lower side of the thumb (enabling them to roost on the smooth inner surface of furled leaves). The pads work by muscular control and wet adhesion (similar to the sucker-footed bats of Madagascar, but quite independently developed). They tend to roost with the head uppermost. Otherwise, apart from minor modifications to the feet and wings, these are externally rather unremarkable bats. Insectivorous with most prey taken by gleaning, hence including a high proportion of non-flying prey including spiders and insect larvae. They roost inside the furled leaves of plants such as banana (*Musa*) and *Heliconia*, where they must change roost site almost daily as the leaves unfurl. Occasionally found amongst dead leaves and palm fronds. Colonies of one to ten bats, not always all in one leaf, but communicating with occupants of nearby leaves.

FAMILY: Mormoopidae.
Ghost-faced or moustached bats. 2 genera, 18
species. Central and South America, just into North
America. Insectivorous.

The ghost-faced bats (*Mormoops* – two species) occur from Southwestern USA south to Nicaragua and from southern Panama into northern South America and through much of the Caribbean. The moustached bats (*Pteronotus*) from Mexico south to northern Bolivia and central Brazil, and parts of the Caribbean. Small to medium-sized bats, 3–27 g (0.1–0.95 oz). The ghost-faced bats have a rather flattened face with round flat ears that are joined across the top of the head by a flap of skin; at their lower point they join a series of flaps and folds under the lower lip. This gives the whole face a rather dish-like appearance, from which the nostrils protrude on a rather rudimentary noseleaf. The *Pteronotus* species have a similar basic pattern, only much less exaggerated, with less-rounded and well-separated ears, and more developed flanges on the upper lip. The eyes are tiny and the wings join the body high up the flank, and indeed join in the middle of the back in two species of 'naked-backed' bats. A section of free tail projects from partway along the tail membrane when the bat is at rest. Insectivorous, ghost-faced bats largely take moths, whereas most *Pteronotus* are more generalist feeders and even take some non-flying prey such as spiders. They mainly roost in 'hot' caves. The colonies are usually large, sometimes to hundreds of thousands and often in mixed-species assemblages with other mormoopids and/or with a wide range of other bat species. Such aggregations can reach more than half a million bats.

FAMILY: Phyllostomidae.
New World spear-nosed bats. 60 genera, 217 species. Southern North America south to *c.*35°S in Argentina and Chile. Insectivorous, fruit- and flower-feeding, a few species carnivorous or sanguinivorous.

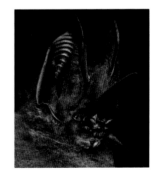

The second largest family of bats and the most physically and behaviourally diverse. Found from the southwest border areas of Texas to California (USA) through Central America and the Caribbean and south through South America to approximately Buenos Aires (Argentina) and Santiago (Chile). Small to large, weighing 5–200 g (0.17–7 oz). Most species have a distinct upright spear-like noseleaf incorporating the nares. In various groups within the family (including fruit- or flower-feeders, insect-feeders and blood-feeders) the face is flattened to some degree and the noseleaf becomes obscure. The tail and tail membrane are well-developed in some species and more-or-less reduced or absent in others (especially in fruit- and flower-feeders). The majority of species (166) are fruit- and/or flower-feeders, 48 species are mainly insectivorous with a few being partially carnivorous and/or frugivorous, and three species are blood-feeders. Roosts known from a wide range of situations, including vegetation (e.g. the tent-making bats), tree buttresses and hollows, caves and various underground or semi-underground artificial structures, and buildings. Some species are solitary or live in small family groups, some in groups with a dominant male and a harem of females, some in large colonies, especially in caves. Some colonies share the roost with other species.

SUPERFAMILY: Vespertilionoidea
FAMILY: Natalidae.
Funnel-eared bats. 3 genera, 12 species. Caribbean, Mexico to Brazil. Insectivorous.

Most species are located in the Caribbean, with four mainland species distributed from northern Mexico to Brazil (São Paulo state). Small, slim species, 2–12 g (0.07–0.42 oz), thus one species, Gervais's funnel-eared bat is another contender for the smallest bat in the world. There are no facial adornments, the ears are quite large and funnel-shaped, the eyes are very small. Male funnel-eared bats have a strange bulbous 'natalid' organ just beneath the skin of the forehead, a large mass of cells resembling sensory cells but of unknown function. Very long legs and tail, associated with broad flight membranes. Insectivorous, taking a range of small, mainly soft-bodied insects and spiders. They roost in warm caves, with occasional individuals sometimes being found in nearby hollow trees or artificial structures. Colonies from tens to several thousand with the individuals spaced and rarely in direct contact with neighbours.

FAMILY: Molossidae.
Free-tailed bats. 22 genera, 126 species. To *c*.40–45°N from southern France in the west to Korea and Japan in the east and similar latitudes in North America, south to throughout South America to *c*.45°S. Insectivorous.

Another large family found through both the Old World and the New World from the mid-latitudes of North America and southern France to most of South America and all of Africa and Australia. Large size range from *c*.3 g to 200 g (*c*.0.1–7 oz). Generally robust bats and very agile crawlers. The face has a large muzzle with jowly lips, often wrinkled. The ears are wide and usually pointing forwards, sometimes part attached along the crown of the head and often connected by a bridge of tissue across the top of the head. The eyes are relatively large. The fur is usually very short, sleek and greasy (two species have very sparse hair and appear more-or-less bald). The tail protrudes for about half its length from the edge of the tail membrane. Insectivorous, generally aerial hawkers foraging in rapid flight, often taking advantage of aggregations or migrating swarms of insects. Roosts often in caves, but also buildings and other artificial structures, as well as rock crevices and tree holes. The blunt-eared bat (*Tomopeas*) roosts alone, but this is rather an exceptional free-tailed bat in many ways, and almost all the other species have communal roosts. Indeed, bats of this family form the largest colonies, with up to 20 million historically recorded from one cave roost site, and there are many other spectacular cave colonies numbering in the millions.

FAMILY: Miniopteridae.
Bent-winged or long-fingered bats. 1 genus, 38 species. Old World tropics and subtropics, north to southern Europe and Japan, south to South Africa and south Australia. Insectivorous.

Occurs through the Old World tropics and subtropics, north to southern Europe and Japan, south to South Africa and southern Australia. Until recently included in the Vespertilionidae. Also of note is that more than half the species have been recognized within the last 20 years, including a single species formerly recognized as the most widely distributed bat (occurring from southern Europe across to Japan and south through Africa and to Australia) now split into a number of species. Small to medium-sized bats weighing 5–22 g (0.17–0.77 oz). These bats are characterized by a plain unadorned face with a high domed forehead and short rounded ears. They have long, narrow wings, with a markedly elongate third finger, such that, when at rest, the wing tip (the first phalanx) is folded back on itself and the second phalanx folded forward again to allow the wing to lie alongside the body of the bat. They are very agile scramblers around the roost. Otherwise a rather unremarkable and uniform group of bats. Insectivorous, preying on a wide range of insects captured in flight, especially moths, beetles and flies. Almost exclusively cave-roosting (or roost in artificial constructions that are similar to caves, such as mines and tunnels) and usually in colonies of hundreds, although often larger, with up to 200,000 recorded. Occasionally small groups or individuals will roost in other structures such as crevices in bridges or in tree holes.

FAMILY: Cistugidae.
Wing-gland bats. 1 genus, 2 species. Africa.
Insectivorous.

Two very closely related species from southern Africa. Until very recently included in the Vespertilionidae, indeed for a very long time regarded as a subgenus of *Myotis*. Small bats, weighing 3–9 g (0.1–0.3 oz). Rather plain-looking bats, but characterized by oval or elongate wing glands located near the forearm and fifth finger in the plagiopatagium. These glands occur in males and females and their number (1–3), exact location and size and shape can vary between animals, and even between the wings of one individual. The function of these glands is unknown, but they are not of the same nature as the scent-producing glands found in the wings (and elsewhere) of other bats. Insectivorous, feeding largely on a range of small soft-bodied insects, mainly flies and hemipterans. Roosts in rock crevices and occasionally buildings, including a church. Mixed sex groups of up to 40 have been found.

FAMILY: Vespertilionidae.
Plain-nosed or vesper bats. 54 genera, 496 species. Worldwide, north to the Arctic Circle, south to southernmost South America, Africa and Australia and New Zealand. Insectivorous.

The largest and most widely distributed family, being found from north of the Arctic Circle in Eurasia to southernmost South America, Africa, Australia and New Zealand, as well as on many remote islands such as the Galapagos, Hawaii and in the southwest Pacific. Mainly small to medium-sized bats, weighing 2–90 g (0.07–3.2 oz), with very few species at the larger end of the range. Generally quite a short muzzle without any facial adornments: in species of *Murina* the nostrils are prominent and tubular, pointing outwards; in a few other genera (such as *Corynorhinus* and *Plecotus*) there are distinct bulbous lobes behind the nostrils; and species of *Nyctophilus* have a well-developed ridge behind the nares, which are on small flaps of skin. The ears are short or long and usually separate but can be joined by a flap of skin across the top of the head. Fur and wing patterning are rare, with most species fairly dull coloured. The tail and tail membrane are well developed, with the tail sometimes extending slightly beyond the edge of the membrane. They are almost exclusively insectivorous and use a wide range of foraging tactics to take all kinds of insects and other terrestrial invertebrates. A few species will take marine crustaceans; fish are preferentially taken by one species (*Myotis vivesi*) and occasionally by a few others, and small migrant birds are frequently taken by at least two species of *Nyctalus* and by *Ia io*. Plant-feeding is only well recorded in one genus (*Antrozous*). They roost in the widest range of possible roost sites, but mainly in caves and rock crevices and tree hollows and increasingly in artificial alternatives such as mines and tunnels, and buildings. Especially vegetation-roosting bats may be solitary or gather in very small groups; maternity colonies more frequently comprise up to 50 or so and can reach hundreds of thousands; hibernation aggregations can also reach 100,000.

CHAPTER 9

Bats and humans

The father of our system for naming and classifying animals, Carl von Linné (Linnaeus), put the bats close to the primates on the basis of some rather superficial similarities. Perhaps it is this similarity coupled with the different and secretive lifestyle that has given bats a somewhat anomalous position in public perception. But while the contribution that bats make to humans and the environment is increasingly being recognized, their role in disease transmission has come under increasing scrutiny, and with the availability of modern techniques is fast catching up with the more traditional studies of rodents and artiodactyls.

Mythology and folklore

Bats have long held a special place in our mythology and folklore, perhaps helped by the uncertainties about what bats are and how they fit into the grand scheme of life. Hence one of the most widespread tales, and one included in *Aesop's Fables* of about 600 BC, talks about competition between the mammals and the birds and the bat always playing along with the winning side. But eventually the birds and mammals both turn against the bats and banish them to live in the night. This basic story, with minor modifications here and there, is remarkably widespread around the world and in a wide variety of cultures, from Europe as far as the aboriginal mythology of Australia and Fiji.

Jens Rydell, and his colleague Johan Eklof, have recently published an account of the part bats play, or played, in Swedish folklore. Their account demonstrates an extraordinary range of powers that were attributed to bats, some for good, some for evil, and many applied to maintain the everyday life of people from

The guano from the wrinkle-lipped bats, *Chaerephon plicatus*, emerging from this cave in Thailand provide an income to the monks that own the cave.

peasants to royalty. Thus bats, or parts of bats (especially such parts as eyes, blood and wings), were widely used to bring success to hunting and fishing, agriculture, house-building, medicine, love and even war. Equally, such measures can be used to ensure that the endeavours of one's enemies and rivals fail. Witches could use bats to harm, and perhaps kill, or to restore life to the dead. Many of the stories here will be familiar to people in other parts of Europe and beyond. Probably very few, if any, of these measures are still practised, but who knows! And certainly, the nailing of a bat to a barn door to ward off evil has been a fairly recent practice.

This great confidence in the powers of bats is somewhat at odds with the distrust that is widespread now – a dislike of something that is not understood or familiar. There can be a general disrespect of bats, as reflected in some of our western expressions such as someone having 'bats in the belfry', or being 'blind as a bat' or just 'batty', or in the Fijian expressions for people who 'smell like a bat', for 'a lazy person who loafs around like a bat' or someone who 'eats like a bat'. The ancient Mayans of Central America had more extreme views of bats,

The Mayans' bat god Camazotz, inhabiting the kingdom of darkness, was greatly feared and widely depicted with similarity to a New World spear-nosed bat.

In Chinese tradition, the figure of a bat represents happiness, and the figure of five bats around a peach tree (the 'Wu-fu') symbolizes the five great happinesses.

associating these animals with death and darkness, and the bat god or death bat 'Camazotz' ('zotz' being the Mayan for bat) inhabiting the kingdom of darkness (the 'Underworld') and decapitating his victims. Thus bats were an object of great fear, and the bat god was widely depicted on urns, glyphs and elsewhere, with the head very frequently bearing a striking resemblance to the head of a phyllostomid bat.

This is in sharp contrast with the Chinese tradition, in which the characters for luck and for bat are both pronounced 'fu'. Thus, the figure of a bat came to represent good luck and happiness and was often incorporated, sometimes in very stylized form, into a very wide range of designs for art, clothing, jewellery, and useful or decorative artefacts. The figure of five bats surrounding the representation of a peach tree is called 'Wu-fu' and symbolizes the five great happinesses of health, wealth, good luck, long life and tranquillity (or a 'good or peaceful death'). Unfortunately, this cultural significance of bats in China seems to have lost a lot of its credence, since a recent survey (2021) of people's attitudes to bats did not show such high levels of reverence.

Bats as totems also feature widely in Europe in such situations as the coat of arms for many European families and in the heraldic crests of many UK families. In the UK several squadrons of the Royal Air Force also include bats in their badges and may be reflected in mottos such as that for the No. 153 Squadron, *Noctividus* or 'Seeing by night', or that for the No. 9 Squadron, *Per Noctem Volamus* or 'We fly through the night'.

Costs and benefits

First we will look at the costs and benefits to bats of human activities. Throughout the world, we have provided a lot of roosts in the form of buildings and in extensive underground tunnel systems. In the buildings we have provided the wood and stone that the bats would naturally choose for roost sites. Depending on the buildings, they can also find the required temperatures and humidity to form their maternity colonies and rear their young. The tunnel systems may be created for human accommodation, mineral extraction, wartime activities, religious functions and a wide range of other functions, and left relatively undisturbed can provide suitable sites for hibernation or for breeding colony use. With this has come massive change in landscape and ground cover, which has brought about major shifts in the balance of bat species. Thus in temperate Europe the reduction and fragmentation of woodland has benefited ubiquitous woodland-edge feeders such as pipistrelle bats (*Pipistrellus* spp.), at the expense of a diversity of woodland specialists, such as Bechstein's bat (*Myotis bechsteinii*). In some cases, we have encouraged the bats by providing food through our agriculture, such as plant products for fruit- and flower-feeders, insect crop pests for a range of species, and concentrations of cattle for vampire bats. In some cases this food production is too seasonal to allow a bat population to take full advantage of it; in other cases it causes conflict with the farmers, particularly with fruit producers who accuse the bats of eating or damaging fruit crops, as well as ranchers in Central and South America, where vampire bats may prey on stock animals. Roosting in buildings, too, has very often brought bats too close to humans where they are unwelcome and measures are taken against them. And we will look later in this chapter at problems associated with bats being forced into a closer and closer cohabitation with humans.

On the other hand, bats are of tremendous benefit to humans. Bat research has revealed discoveries of the mechanisms of such things as thermoregulation, blood anticoagulants, flight mechanics and sonar, and bats provide a massive

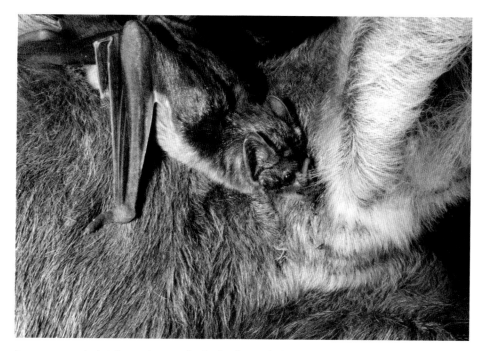

A common vampire bat, *Desmodus rotundus*, feeding from a donkey. Study of vampire bats has been of great benefit to our own medical developments.

contribution to ecosystem services in the management of pests, and the pollination and seed dispersal of a wide range of plants that might be important for timber or for food and drink, or even more generally to prevent erosion of coastal and hilly regions and at the individual level often to provide useful shade trees.

The recognition that bats can be useful for insect pest control was perhaps first identified by Dr Charles A. R. Campbell, in the early years of the 20th century. As a Public Health official in Texas at the time that it was shown that malaria is transmitted by mosquitoes, he was looking for a means of mosquito control and decided that bats should be encouraged. He dedicated himself to constructing large towers to house the bats and eventually did achieve some success in attracting many in. And he claimed in his fascinating book *Bats, Mosquitoes and Dollars* (1925) that examination of droppings showed the bats to be taking a lot of mosquitoes, and there was a fall in the incidence of malaria in the area. Meanwhile he had given up his job and eked out a living from the sale of guano from 'Dr Campbell's malaria-eradicating, guano-producing bat roost'. Local municipalities took up the idea and built bat towers, and the idea was

A Campbell-style bat tower as a roost for large numbers of bats as a means of controlling insect pests.

picked up abroad. Bat towers to Campbell's plan were built in The Netherlands and Italy, but they didn't have the right bats to work; however, one in Malaysia was reckoned to be effective in mosquito control, as was one of a different design in South Africa, which was installed to reduce a high incidence of malaria in railway workers – this tower still stands and, despite the need for some repair, still houses bats.

On a much smaller scale, bat houses are being introduced around a range of crops, where it is hoped that bats will provide pest management control services, and for some of these crops, such as grape vines and rice fields in Portugal, the bats are proving beneficial. Portugal also has a couple of key 18th century libraries, where bats are encouraged to roost in the library – the librarians have long felt that the bats feed on the insects that might otherwise be feeding on the books and manuscripts.

A further development of recent years is research to provide hard evidence of the financial value of the contribution bats make to pest control of certain crops. Thus, one calculation suggested that the predation of cotton bollworm moths by Brazilian free-tailed bats in Texas provided pest suppression services to the value of about $650,000 per year, about $100,000 of which is accounted for by reduced pesticide use. Another study suggested that the pesticide replacement costs might be valued at $500,000 per year. A project in one region of Mexico suggested that bats might be responsible for reducing crop damage by 25–50 per cent or between $500,000 and $1.2 million per year. Overall, it has been calculated that pest control by bats saves US farmers $23–53 billion dollars annually. The bats from one cave in Thailand are estimated to save Thai farmers about $300,000 per year through control of rice pests – and one large bat cave can produce guano valued at $100,000 per year as fertilizer. The value of bats to pollination and seed dispersal is much more difficult to calculate, and it tends to be the damage the bats do that gets noticed. Nevertheless, over 1,000 plant species are recorded from the diet of Old World fruit bats, mainly from fruit-feeders, with less study of flower-feeders.

The larger colonies of bats can be of great benefit for insect pest control, pollination and seed dispersal, and production of guano as fertilizer.

Some of these plants are solely, and others chiefly or significantly, pollinated or seed-dispersed by bats, including plants that provide timber or other wood products, food and drink (including fresh fruit), medicines, dyes and fibres and some ornamental plants. Similarly, for the New World spear-nosed bats over 500 plant species are recorded from diet analysis, including several hundred pollen and nectar providers, and again, many of these are trees and bushes that are important to humans in a wide variety of ways.

Other value that the bats bring can be found in, for example, ecotourism. Every evening about 50 people gather to watch the emergence of the 5 million wrinkle-lipped bats from Deer Cave in Gunung Mulu National Park, Sarawak. Similarly, about 100,000 tourists each year gather to watch the 1.5 million Brazilian free-tailed bats emerging from their roost beneath Congress Avenue Bridge in Austin, Texas, and similar numbers gather at Carlsbad Caverns, New Mexico, for a similar spectacle. Many people flock to Kasanka National Park in Zambia in November and December to watch the evening dispersal of 5–10 million straw-coloured fruit bats that temporarily collect there.

The daily evening gathering of people in Sarawak watch the emergence of a large colony of bats from the nearby cave.

And one further use that bats were put to was as (potential) weapons of war. Jack Couffer, in his book *Bat Bombs: World War II's Other Secret Weapon*, relates how he was one of a team of researchers engaged on a US project to attach incendiary devices to millions of bats that would be released over Japanese cities to 'frighten, demoralize and excite the prejudices of the people of the Japanese Empire'. A lot of serious original research went into the project, and considerable ingenuity to develop the devices, their harness and triggering mechanisms, as well as their means of transport and release. And it might have worked; indeed, in one demonstration some 'armed' bats escaped and completely destroyed the control tower, barracks and other buildings of a newly constructed airfield. But in February 1944 orders came to discontinue 'Project X-ray' on grounds that did not ring true with the team, which feeling was perhaps confirmed by subsequent events.

Zoonoses

One developing topic of concern is the association, or presumed association, of bats with diseases transmitted to humans (zoonoses). One such disease of minor importance is lung infection by the fungus *Histoplasma capsulatum*. *Histoplasma* exists in bat caves, but it is equally common in bird colonies or just free-living in soil. It rarely causes serious problems but can occasionally lead to lung lesions. It is something that cavers, in particular, remain aware of, and many show positive in skin tests, although they very rarely develop symptoms. There is a range of bacteria, such as *Bartonella* and *Campylobacter*, which can be a source of illness in humans, but many are associated with a wide range of wildlife and may require a vector to perform transmission. None of these is of major concern.

Of more concern are some of the viruses, and a number of recently 'emerging diseases' have been attributed to being of bat origin. Not so much an emerging disease, but rabies kills over 50,000 people per year, almost all in Africa and Asia and almost all from the standard rabies virus, RABV, and transmitted by dog bites. This *Lyssavirus* can also occur in vampire bats in Central and South America and causes human infections from time to time. A number of strains of this virus are identifiable in bats in North America, where there are occasional records of human deaths (perhaps one per year) being associated with one or two strains, such as that of the silver-haired bat, *Lasionycteris noctivagans*. Almost 20 other *Lyssavirus* species (rabies-related viruses) are currently recognized, most described within the last 20 years following great advances in the diagnosis of virus types. Most of these

species have not been associated with human infections, but human deaths have been attributed to Duvenhage (South Africa, one case), Australian Bat Lyssavirus (two cases), European Bat Lyssavirus 1 (three cases) and European Bat Lyssavirus 2 (two cases). EBLV1 is quite common throughout Europe in serotine bats (*Eptesicus serotinus*), a common house bat, but the number of human infections is extremely low. The host for EBLV2 is the much less common and less synanthropic (living close to humans) Daubenton's and pond bats (*Myotis daubentonii* and *M. dasycneme*). Most bat workers are vaccinated, and current vaccines are considered protective against most lyssaviruses, but possibly not for one or two of the viruses more distantly related to RABV. In many parts of the world there are surveillance schemes in operation to monitor the occurrence of lyssaviruses.

Until recently, rabies was the only viral disease considered closely associated with bats. Since then a number of other zoonoses have emerged and have been identified as having their evolutionary origin in bats.

A coronavirus, severe acute respiratory syndrome, or SARS (transmitted by the virus SARS-CoV), appeared at the end of 2002 and rapidly spread to about 28 countries, infected *c.*8,000 humans and killed *c.*800. The virus was isolated from palm civets and raccoon dogs in Chinese 'wet' markets, and very similar viruses were isolated from a variety of horseshoe bat species, in particular *Rhinolophus affinis*, as well as a bent-winged bat and a rousette fruit bat. Horseshoe bats are not common in these markets (which were believed to be the source of infection) and the viruses found in the bats are not identical, so while it seems likely that the virus is of bat origin, it is considered that it probably moved through an unidentified intermediate host, but when that spillover occurred remains unknown.

Ten years later another conoravirus caused human fatalities: Middle East respiratory syndrome (MERS) was identified from the Arabian Peninsula in 2012. It too, spread around the world, mainly by human-to-human transmission, and has infected about 2,500 people, of which *c.*900 have died. To date, despite continued sampling of bats, only one partial fragment of RNA of the MERS virus has been found in a bat – an Egyptian tomb bat, *Taphozous perforatus*. Meanwhile, many human infections have been traced to contact with camels, which show a high incidence of the virus. Again, if this virus is of bat origin, it is not known when the spillover to camels occurred and why it suddenly emerged as a problem.

The silver-haired bat, *Lasionycteris noctivagans*, is a North American species associated with occasional rabies infections in humans.

The emergence of these two diseases encouraged a much-enhanced interest in coronaviruses in bats and they were found to be widespread and often quite host-specific. That research has proved useful in investigations of a third coronavirus global pandemic, which emerged at the end of 2019 and continues at the time of writing. Closely related to the original SARS virus, this one is named SARS-CoV-2, and the disease it spreads is COVID-19. A similar coronavirus probably evolved in bats many years ago and later jumped or 'spilled over' to another animal species, as yet unknown, where it mutated to the virus that now infects humans. SARS-CoV-2 has never been found in bats, and infection with this virus is exclusively from person to person.

At the time of writing there have been over 120 million confirmed cases of COVID-19, with more than 2.5 million deaths in 192 countries or regions. The possibility of a 'wet' market being the source via overspill through a pangolin or a snake has been proposed but widely discounted, as has the virus escaping from a laboratory, which has been dismissed by a World Health Organization investigation. The search continues for the source and mechanisms for infection, and for treatment. Also at the time of writing, concern has been expressed about the possibility of this virus being transmitted by infected humans to wildlife, including bats. That could have a severe impact on the bats themselves but also on public health management. Currently, restrictive guidelines are in place to limit the risk of scientists, cavers or others transmitting the virus to wildlife.

Paramyxoviridae is another family of viruses that includes species involved in zoonoses of bat origin. Hendra virus was isolated in 1994 in Australia (as equine morbillivirus) with outbreaks of an acute respiratory disease that affected about 26 horses, which, in turn, affected four humans; two of the human infections were fatal. It was subsequently established that the virus occurred naturally in all four species of flying fox (*Pteropus* spp.) in Australia and could be passed in urine. In all there were seven human cases, including two more human deaths, but awareness and appropriate husbandry of horses and other health provisions have so far proved effective for disease control.

The closely related Nipah virus was described in 1999 in Malaysia following a disease outbreak in pigs and humans. In all, 105 of 265 human cases were fatal in the initial outbreaks, with infection being contracted from infected pigs. Again, flying foxes proved to be a natural host and reservoir for the virus. One of the proposed hypotheses for the emergence of Nipah virus in Malaysia

The Indian flying fox, *Pteropus giganteus*, has been associated with outbreaks of Nipah virus in South Asia.

was the effect of an El Niño (climate pattern) event on the movements of the flying foxes, which had forced them to move from traditional foraging areas where food supplies had been damaged; whether there is any truth in that, it was unusual to have a *Pteropus* colony roosting above several thousand pigs. Again, the introduction of appropriate farm management practice and outbreak management measures has kept outbreaks under control. Direct infection of humans from *Pteropus giganteus* has been shown in seasonal, geographically scattered outbreaks in Bangladesh, where fresh bat secretions (mainly urine) contaminate freshly tapped palm syrup drunk by the local peoples. Efforts to prevent access by bats to the syrup pots, by covering them with bamboo skirts, has reduced infection rates. An outbreak of Nipah virus in Kerala, South India, in 2018 claimed 17 lives. It was believed to have been contained and was declared over in about a month, but there was one more death reported in 2018. A number of other Paramyxoviridae, including Menangle and Tioman, have been associated with bats, but not yet with any zoonosis.

The Filoviridae includes the viruses responsible for Ebola and Marburg. Marburg, closely related to Ebola, has occasional outbreaks such as in mine workers in Uganda and even cave tourists. An outbreak in Angola in 2006 resulted in 252 cases, most of which were fatal. Bats have been implicated both epidemiologically (human cases have been associated with caves), and both serological and viral-RNA-positive *Rousettus* have been found, which fundamentally supports the theory that bats are reservoirs of these infections.

However, despite intensive field research on a wide range of organisms, from plants to bats and birds, a natural reservoir of Ebola viruses has not been identified. Laboratory evidence that bats (the fruit bat *Epomophorus wahlbergi* and *Tadarida* spp.) could survive inoculation with Ebola virus suggested that bats might be a natural reservoir. Nevertheless, fieldwork, including the sampling of sometimes large numbers of bats (e.g. 539 of 18 spp., including large numbers of 'Tadarida' spp., after an outbreak at Kitwit, Democratic Republic of the Congo), has not been able to confirm this (although it is still suggested that bats were the most likely reservoir). However, another almost contemporaneous report provides evidence of infection by Ebola virus (i.e. the presence of antibodies) in three species of fruit bats amongst 679 bats (of 1,030 wild-trapped animals) tested in 2002 and 2003. The fruit bats implicated were *Hypsignathus monstrosus*, *Epomops franqueti* and *Myonycteris torquata*. An alternative proposal about the spread of Ebola, that it has moved in a fairly linear pattern of waves, does not rely on a permanent natural reservoir for the virus (although it is true that even this hypothesis does not rule out bats as a reservoir). The most recent major outbreak infected nearly 30,000 people, of which 11,000 died. Again, there was some evidence that fruit bats might be implicated, but in this outbreak the first victim (the 'index case') was a small boy who played in a hollow tree where *Mops condylurus* roosted. Bushmeat is also likely to be a source of infection for a virus with a very high infectivity.

There is a wide range of other zoonoses or other diseases for which bats have been at some time implicated, such as AIDS, West Nile fever and dengue. For some there is no evidence of bats being involved in transmission; for others their role is so minor as to be negligible. Nevertheless, there is a wide range of viruses associated with bats and evidence that they have an underlying immunity to many of the viruses they carry. Current research suggests it is possible that the evolution of flight in bats has favoured certain genes that encourage antiviral immune responses to control viral propagation and limit the inflammatory responses to the viruses that cause problems in other mammals. Presumably bats have lived with these viruses for a very long time and, perhaps with the exception of Ebola virus, which has been around a long time, it is in these modern times, with humans increasing encroaching into natural habitats and forcing ourselves into more direct contact and conflict with animals, that we see, and will continue to see, the emergence of such zoonoses of unpredictable impact. Further research into the way bats cope with these viruses may help us cope with future epidemics.

CHAPTER 10

Conservation

Under a system created by the International Union for Conservation of Nature (IUCN) animal species are assessed for their conservation status under an agreed set of criteria to produce the Red List of Threatened Species. At a global level, 191 of 1,400 bat species (over 13 per cent) are currently considered to be threatened with extinction. This includes seven species for which all the evidence points to them now being extinct. For the categories of threat, 22 species are in the most threatened category (Critically Endangered), 55 are Endangered and 107 are Vulnerable. Of the others many are so recently recognized as species that there has not been sufficient information to carry out an evaluation of their conservation status. Many others have been evaluated but there is still insufficient information available to categorize them with any confidence, and so they are classified as Data Deficient. That leaves a number of species as Near Threatened, i.e. likely to fall within one of the threat categories in the near future unless circumstances change, and many species that are regarded as not currently threatened with extinction within the near future. Some of the major threats are shared with other animals and in particular include changes to land management resulting in great loss of habitat, other pressure from human populations on the environment, and climate change.

The Old World fruit bats – a special case

The family of Old World fruit bats includes the highest proportion of threatened species (35 per cent). A large number of the 191 species are restricted to an island or a small group of islands, where they exist in populations restricted

A Southeast Asian species of Old World leaf-nosed bat, *Hipposideros diadema*, one of a group of bats with a high proportion of threatened species.

by the island size and available habitat. Six of the seven global extinctions are of flying foxes endemic to small islands. Less than half (46 per cent) of the species are considered as not threatened with extinction at the present time. On many islands the bats have been confined to remnants of natural habitat by development for agriculture or habitation; the alternative is that with the loss of their natural roosting and foraging habitat, the bats have frequently been forced into living in closer proximity (and potential conflict) with humans. Bats, especially fruit bats, are widely considered as good eating (and there may be not much alternative wild meat available on many islands), and in some Pacific islands (e.g. the Marianas) the eating of fruit bats is of high cultural significance. This has led to overexploitation in some islands. On a few of these islands this has been compounded by predation on the fruit bats by the introduced brown tree snake, thought to be a major contributor to the extinction of the endemic fruit bat of Guam, *Pteropus tokudae*. On Christmas Island (Indian Ocean) the introduced crazy ant (*Anoplolepis gracilipes*) is thought to have been a major factor in the extinction of the Christmas Island pipistrelle bat (*Pipistrellus murrayi*) and declines of the endemic subspecies of the black-eared flying fox (*Pteropus melanotus natalis*) through disturbance to roosting bats and damage to vegetation, including destruction of fruit trees. Very rapid forest removal in the Comores has led to the near extinction of Livingstone's flying fox (*P. livingstonii*).

Cyclones are of regular occurrence in many of the islands of the southwest Indian and Pacific Oceans. These can destroy habitat and reduce food supply with consequent catastrophic population crashes. Thus the population of the flying fox endemic to Rodrigues, a small island close to Mauritius, in the Indian Ocean, was reduced to about 70 individuals in the mid-1970s. This was due to damage to their roosts and foraging habitat and destruction of a large part of their food supply. A few were taken into captivity and subsequent populations maintained in zoos around the world ready for return to the island should that become necessary. Fortunately, while the wild population has had its ups and downs since then, it has overall recovered well and is once again well established on its island, with a population of *c.*20,000. Its neighbour in Mauritius is not faring so well. In Mauritius, there has in recent years been a major move to develop more fruit farming, especially of fruit such as lychees, longans, mangoes

Rodrigues flying fox, *Pteropus rodricensis*, an Indian Ocean island endemic, nearly went extinct in the 1970s; habitat revival and captive breeding ensured its survival.

RIGHT The taking of bats for food, mainly the larger fruit bats, is a major conservation problem in some areas.

BELOW The value of bats to pollination and seed dispersal is widely underestimated.

and guavas. While repeated research shows that fruit bats are not a major source of damage to such fruit crops, the fruit farmers have persuaded the government to initiate a number of culls based on unrealistically high population estimates and so place the species in a threatened category. In The Maldives, to the north of the Indian Ocean, the endemic subspecies of *Pteropus giganteus* has been regarded as a pest of fruit crops and the islanders have exercised a wide range of control measures against the small and sparse bat populations that occupy many of the small islands that make up The Maldives. The government sought further advice and was advised that a cull of up to 49 per cent per year would maintain a stable population. This would actually result in a very high risk of extinction of the bats in quite a short period, and so efforts have been made to address the problems in other ways. Apart from the successful recovery of the Rodrigues flying fox, the Pemba flying fox (*P. voeltzkowi*) also made a successful recovery from a low of a few hundred animals in a couple of roosts to a population of nearly 30,000 in over 40 roosts. This recovery was largely due to educational programmes and roost protection by co-ordinated community groups.

On the mainland, many of the same problems apply: there is the loss of natural habitat and the bats choosing to, or being forced to, move more into artificial habitats, and into competition with fruit growers both at the commercial scale and in backyard economies. In the former, there are means of fruit protection (such as fixed netting), but these involve a capital outlay and maintenance costs that the farmers do not want; nevertheless, whether it is bats, birds or other mammals that are causing most of the crop damage, that may be the best long-term solution. Bats as bushmeat is also an important issue in mainland areas, such as West Africa, and the taking of bats for food is often undertaken on a commercial scale, especially from the larger colonies in trees or in caves. The taking of bats for the restaurant trade is also of concern in some areas where it is believed to be damaging populations of fruit bats, especially island fruit bats, such as formerly on the Seychelles, and currently in the South Pacific and Madagascar. On the other hand, where large colonies of bats are living in some temple caves in parts of Southeast Asia, they are not disturbed, and indeed are well protected by the monks responsible for the cave.

Another thorny issue is any role the bats have in the maintenance and transmission of diseases of animals and humans (see p.131). But much more important than the actual involvement in disease transmission is the perceived role that obvious colonies in trees and caves may have, making them easy targets for campaigns against them. The same applies to the situations in which bats are just not welcome, for whatever reason. Offset against the situations in which fruit bats are seen as a problem are the benefits they bring in the pollination and seed dispersal of a wide range of plants, both natural and cultivated, which are useful to us in a variety of ways, including for building, food, medicine, shelter and in the maintenance and recovery of natural habitats.

Tourism is a mixed blessing for colonial fruit bats. It is good for people to be able to see and appreciate these animals in colonies in trees or caves, but it is easy for this to cause excessive disturbance to the bats, especially in caves, unless carefully controlled. When grey-headed flying foxes moved into the Gordon suburb of Sydney, they were very unwelcome and there were strong calls to get rid of them. However, some of the local residents decided to make a feature of the bats, and following the construction of viewing platforms and creation of an education programme (such as information boards and guides), the bats were much better appreciated and the calls for their removal declined. The same species was, however, a problem in Melbourne's botanic gardens, where their growing presence (in time and number) was causing damage to specimen trees. Not an easy thing to do, but the bats were eventually encouraged to move their roosts to another area outside the city and its botanic gardens. In the last ten years a colony of c.20,000 of the same species has established in South Australia, in Adelaide, where the residents are uncertain as to how welcome they are. Elsewhere, visits to large colonies of fruit bats are well established on the ecotourism trail, such as those to see the millions of straw-coloured fruit bats that gather annually at Kasanka National Park in Zambia.

The concerns about mass mortality events in bats has been mentioned elsewhere (see p.98). There are several such instances with fruit bats, such as mass mortality of the Marianas flying fox on the Mariana Islands in the South Pacific, where the mortality coincided with a major outbreak of measles in the human population. In Australia, there have been recent instances of large numbers of flying foxes killed through excessive heat (heat waves of up to 40°C) with many being rescued by bat carers – and such events are becoming more frequent.

The other bats

While the fruit bats have their special issues, they also share problems with many of the other bats. However, it seems that the family of New World fruit- and flower-feeders of the spear-nosed bats (Phyllostomidae) does not suffer the same level of threat; only 14 (6.5 per cent) of species are included in threatened

This Schneider's leaf-nosed bat, *Hipposideros speoris*, is
safe in a well-protected temple cave in Karnataka, India.

categories of the Red List. With these bats there is not the reliance on limited
resources of islands and, despite rapidly increasing pressure, there is still enough
suitable habitat on the large mainland areas to maintain most species. It is also
likely that there is insufficient evidence for a lot of species that might well qualify
for inclusion in categories of threat.

One group that does seem to suffer a high level of threat is the horseshoe
bats and their allies (the Old World leaf-nosed bats and trident bats). For this
group, 30 species (15 per cent) are considered threatened. The threats may be
that these bats seem particularly sensitive to disturbance and are essentially cave
bats that are subject to disturbance from mineral exploitation, caving and cave
tourist pressure, guano collection for fertilizer or the exploitation of swiflet nests
for birds' nest soup, where these activities are not well managed. In addition, there
is the deliberate targeting of bats in caves for eradication associated with fruit
protection or disease management (there was a marked increase in attempts to
destroy bat colonies in various parts of the world following the outbreak of SARS-
CoV-2), even the closure with inappropriate materials of caves for 'safety' reasons

or to protect important artefacts. Most of these bats are primarily woodland bats and the rapid decline of native forest is a widespread threat. Many horseshoe and leaf-nosed bats have moved into buildings, where they may be persecuted or unintentionally affected by destruction, refurbishment or pesticide use. The two most widespread European horseshoe bats, the greater horseshoe (*Rhinolophus ferrumequinum*) and lesser horseshoe bat (*R. hipposideros*), both underwent huge declines in the first half of the 20th century, and most markedly in northwest Europe. Giving the species a high level of protection and investing a great deal of conservation effort and money into the protection and enhancement of roosts, and engaging landowners in the appropriate management of suitable foraging areas, has brought encouraging signs of recovery in some areas.

Another highly threatened family is the Mystacinidae of New Zealand, but there are only two species. The greater short-tailed bat, *Mystacina robusta*, was recorded as quite abundant on Big South Cape Island and one or two of its neighbouring islands off the southern tip of New Zealand until rats reached the islands in the 1960s and became major predators and competitors. The last confirmed sighting

In summer, tourists gather on Congress Avenue Bridge, Austin, Texas, to watch the evening emergence of up to 1.5 million Mexican free-tailed bats, *Tadarida brasiliensis*.

DAMAGE TO HABITATS

As well as the special problems bats face through their communal roosting behaviour, they face widespread conservation problems of damage to habitats and the landscape continuity needed for their movement between feeding grounds and on migration. Globally the most important habitats for bats are forests and wetlands, which is where the main diversity is, although a wide variety of bats exist in other habitats, including a surprising array of species, both insectivorous and plant-feeders, that live in arid zones. Increasingly, landscape exploitation from logging, development, agriculture and the use of pesticides (which can affect bats directly or through their food supply) is affecting the conservation status of many bat species. Fires, even in areas where fires have been a natural process, are increasingly a problem, with longer-term and more extensive damage to the associated habitats. There are, of course, a few species that have adapted well to the human-created habitats, finding suitable roost sites in associated artificial structures and a rich source of insect food from crop pests, where the insect supply can be sustained throughout the bats' active period.

Logging and forest clearance are a major concern; even selective logging can result in high levels of damage to the remaining forest and probably removes the trees most likely to have cavities suitable for bats. Forest clearance may make way for land for ranching or for crop production. Even where natural forest is replaced by other trees, these are often introduced species that do not have the diversity of associated wildlife. Niah National Park in Sarawak, Borneo, is an important reserve, with its caves formerly hosting c.200,000 naked bats with a range of other bat species. The reserve is now a small island in a sea of oil palm plantations, and the naked bats have all but disappeared. In a curiosity of national legislation, the earwig that lives on the naked bat is a great national treasure and very highly protected, indeed is probably more highly protected than the bat it relies upon.

Wetland habitats have been similarly lost or damaged. Historically, wetlands have provided a huge diversity and abundance of insects. Drainage and canals, along with pollution and other factors, have drastically affected that food supply. A low level of some pollutants can actually increase the insect biomass, but at the expense of the diversity of prey, and possibly only for a short period of the year. The European Daubenton's bat, which feeds on emerging aquatic insects, can take advantage of a low level of pollution to feed on the large volume of one or two species of chironomid midges at certain times of year, but that water will not provide for its needs for much of the rest of the year and will not provide for the diversity of bats that may be seeking the food provided by a more diverse insect fauna.

The Niah Caves National Park, Borneo, is almost completely surrounded by oil palm plantations.

of this bat was in 1967, five years after the species was first described, and for a long time it was considered extinct. Big South Cape islands and one or two of its neighbours have been subjected to rat eradication programmes, and an echolocation call that could be this bat species was recorded on Putauhina Island in 1999. Subsequently there has been sightings of bats here and on Big South Cape Island, but as yet there is no confirmation that this species is extant, and it is regarded as Critically Endangered. The lesser short-tailed bat, *M. tuberculata*, is only found in New Zealand, where it is widely distributed on the mainland, and maybe more abundant than the only other New Zealand bat, the long-tailed bat, *Chalinolobus tuberculatus*. Nevertheless, its population is greatly reduced and is now concentrated into a limited numbers of regions, mainly through predation and deforestation. The high level of foraging on the ground by this species puts it particularly at risk of predation by a range of introduced mammals, including rats, cats, stoats and possums. And its terrestrial behaviour also has put it at special risk from pesticide programmes through direct contact or the taking of contaminated prey. These bats collect into large colonies, mainly in trees, where they are also prone to competition with accidentally introduced wasps. For these reasons and the fact that it is in a family of its own, it is regarded as Vulnerable. And so too, then, is the very curious and distinctive wingless fly, *Mystacinobia zelandica*, which lives only in close association with this bat (but does not appear to be parasitic) and also belongs to a monotypic family restricted to New Zealand.

There are still a number of conservation issues that are more specific to bats. One is associated with the vampire bats of Central and South America, where they often cause a problem for humans through feeding on the blood of introduced cattle and horses and are from time to time associated with outbreaks of rabies in domestic animals and occasionally humans. Bat attacks on humans are also initiated by people moving into areas, such as to open or reopen mines for mineral exploitation, and clearing out indigenous animals that would have provided food for the vampire bats. These activities have resulted in extermination campaigns against vampire bats, which have had little regard to the correct identification of target colonies and so have resulted in the extermination of large numbers of harmless or beneficial bats.

A new problem for bats is wind turbines. When first introduced these were seen as a minor issue, but it rapidly became apparent that the turbines are killing significant numbers of bats, particularly migratory bats, either by direct collision or through barotrauma (where the bats get close enough to the blades to experience a sudden and precipitous drop in air pressure such that their internal organs rupture). Bats, more than birds, seem to be attracted to the turbines, although why is not fully understood yet; it has been observed that bats from mainland Sweden will fly up to 14 km (9 miles) out into the Baltic Sea to forage on insects accumulated in the warmer air around the individual structures of off-shore turbines. Around the world, bat specialists are working with governments and the industry to try to find ways to minimize the impact of wind turbines through considerations of their siting (especially in relation to the migration patterns of the bats), and the timing and conditions under which the turbines are operating, including special measures associated with the times of year when bats would be most vulnerable. A good example of such collaboration can be seen in the Intersessional Working Group on Wind Turbines and Bat Populations of EUROBATS, which coordinates the findings from throughout Europe and beyond and identifies further needs for data and research.

Another new threat for bats, white-nose syndrome (WNS), was discussed earlier (see p.97). This disease is associated with a fungus and has killed millions of bats in North America and would appear to have been a recent introduction from Europe, where the fungus is widespread but not a problem to bats. With increasing movement of goods and people, it is quite likely that other such problems will arise. Similarly, the emergence of diseases in humans and domestic animals has resulted in campaigns against bats, often with scant regard as to the role of bats in the transmission of such disease. This has been seen recently in campaigns to kill bats following the emergence of SARS-CoV-2 and the ensuing global COVID-19 pandemic, campaigns that have been waged in various areas throughout the world and irrespective of the type of bat, and are usually just focussed on the bat colonies that are the easiest target.

Climate change

As we move into the Anthropocene, the era when humankind has the greatest influence on the changes to a fragile planet, one of the main drivers of change is increasingly considered to be human-induced climate change. For bats this may have an impact on distribution (and probably population levels), as well as the timing, routes and distances of migrations, changes in the phenology of food availability at crucial times, such as at times of emergence from hibernation, times of lactation or weaning of young (all of which may have particular impact on juvenile survival), and the ability of bats to adapt to landscape changes. Equally, the ability of landscape structure to support range shifts and migration corridors that provide suitable foraging habitat, and the availability of suitable hibernation sites, are key factors. Extreme weather events are also a major issue. Also important is our own ability to adapt in order to limit the impact or to enable wildlife to cope with climate change.

Perhaps at particular risk are island bat species and other species with limited opportunity to shift their range, such as high-altitude or boreal species. There are relatively few boreal species, with typical examples being the north European northern bat (*Eptesicus nilssonii*) and parti-coloured bat (*Vespertilio murinus*). Since they cannot move further north, these may face pressures if species occupying similar niches were to move further north and become competitors. A study in Costa Rican cloud forest showed that a number of bat species had moved to higher altitudes to become sympatric with high-altitude specialists, although at the time of the study there was no evidence of negative impacts of this on the high-altitude specialists. Aldabra Atoll is a small island (a land area of *c.*150 km², or 58 square miles) in the western Indian Ocean. It is a raised coral atoll with very little land higher than 3 m (10 ft) above sea level. And it has four species of bat, three of which are endemic to the island: a flying fox, a trident bat and a free-tailed bat (although the last has recently been considered synonymous with populations on the Comoro Islands). All are present in very small populations. Sea level change would severely impact the island and would likely lead to the extinction of all three of these species.

Parti-coloured bat, *Vespertilio murinus*, one of the few species restricted to the far north of Europe and Russia and which may be early to suffer from climate change.

Pacific sheath-tailed bat, *Emballonura semicaudata*, is restricted to the
South Pacific islands, many of which are very low lying and liable to effects
of climate change.

Of course, there are many such low-lying islands including the entire Maldive
Islands archipelago with a maximum elevation of just over 2 m (6.6 ft) above sea
level. In the Pacific, too, there are many such islands that are home to the Pacific
sheath-tailed bat, *Emballonura semicaudata*, which mostly roosts in sea caves and
coastal rock overhangs and would be extremely vulnerable to rises in sea level.

The bat conservationists

Bats are protected in most countries around the world. That may be as part of
general wildlife legislation but is often more specific and may apply conservation
measures to roosts and foraging habitat as well as the animals themselves. Apart
from the broad obligations taken by countries through such conventions as the
Convention on Biological Diversity (CBD; Rio, 1992), there are international
treaties that include measures specifically for bats. The Convention on
International Trade in Endangered Species of Wild Fauna and Flora (CITES)
includes protection for a number of flying fox species in the South Pacific, which

Control of the taking of fruit bats such as this Marianas flying fox, *Pteropus mariannus*, for food and medicine in parts of the Pacific has been helped by their inclusion in CITES.

prohibits the international trade in some species and seeks careful monitoring of such trade in other species. Because of their migratory traits, some bats are protected through the Convention on the Conservation of Migratory Species of Wild Animals (CMS), which is the mother convention for the intergovernmental Agreement on the Conservation of Populations of European Bats (EUROBATS). In its 30 years, 37 states have become Party to EUROBATS, from Portugal to Georgia, and Norway to Malta. This United Nations Environment Programme (UNEP) treaty promotes international collaboration between governments inside and outside the Agreement and with a wider community, and it has developed a joint EUROBATS/ EU bat action plan. Through its Advisory Committee (AC), its Publications Series and other documents provide principles for the development of national protocols for bat conservation activities. The AC also identifies and investigates new issues as they arise, and assesses applications to the EUROBATS Project Initiative, which helps fund projects that meet the aims and interests of the Agreement. EUROBATS also promotes public awareness campaigns, such as the annual European Bat Night. The Agreement has strongly encouraged and supported non-governmental organizations (NGOs), and the networks that have been established among bat workers throughout the range states have been immensely valuable in advancing bat conservation.

In Europe most states have a national NGO concerned for the conservation of bats, a number with a specific organization for bats. These organizations have formed a partnership, BatLife Europe, for action on shared concerns. A joint programme for the conservation of migratory bats was established in the 1990s between North America and Mexico and has grown to become a much bigger organization with broader aims, as the Latin American and Caribbean Bat Conservation Network. There are other such international alliances of NGOs or scientists (some including government support and involvement) for Australasia, Africa, North America, South Asia and Southeast Asia. Bat Conservation International has a global remit for the conservation of bats, and Lubee Bat Conservancy (established by Luis Bacardi of the rum-manufacturing family) has a main focus on the conservation of Old World fruit bats. The IUCN has its Bat Specialist Group, a global network of c.150 bat specialists concerned for bat conservation. Amongst the products of this group have been two action plans, one for the Old World fruit bats published 30 years ago (and under revision) and one for all other bats published 20 years ago. But, as an indication of how global priorities can change, it is interesting to note that neither of these volumes mentions concerns about wind turbines, and there is almost no mention of emerging diseases, either zoonotic (e.g. SARS) or endemic to bats (e.g. WNS), all of which closely followed the second action plan and all of which have had global impact.

As conservation issues are identified, the focus of action may be in research (including survey and monitoring) or education, site protection or the wider protection of habitats, including roosts, foraging habitats or features of the commuting or migratory routes that the bats require. And, as discussed above, very much a growing concern of bat conservationists is the impact of climate change, how to adapt for that, and how bats might be useful indicators for the impacts of such worldwide events. The conservation of bats relies on good knowledge and evidence of the animals and their requirements. And that is one area in which for bats, perhaps above all other mammals, there have been huge advances over the last 50 years. Over this period the advances in techniques to capture bats and study them in the field have allowed detailed studies that had just not been possible previously. This includes the steady development of more and more sophisticated bat detectors that convert the ultrasonic calls of bats to something we can hear, record and analyse, the development of marking and tracking techniques that enable the study of bat behaviour in the field, equipment that allows the remote observation and monitoring of bat behaviour in the roost without disturbance, the ability to take smaller and smaller samples in a non-invasive or marginally invasive way that can be used in the laboratory in a huge range of studies, from taxonomy to diseases. All of these have contributed to a great burgeoning of knowledge and understanding of bats and their conservation needs.

Nevertheless, there is a great deal more research and policy development needed to ensure that we maintain the diversity of bat species, including many that are of great benefit to us and our way of life.

SUMMARY TABLE OF BAT FAMILIES OF THE WORLD

Family	Common name	No. of genera	No. of species	Distribution	Food
Pteropodidae	Old World fruit bats	46 genera	191 species	Old World tropics	Fruit- and flower-feeders
Rhinopomatidae	Mouse-tailed bats	1 genus	6 species	Old World tropics and northern subtropics	Insectivorous.
Craseonycteridae	Hog-nosed bats	1 genus	1 species	Thailand and Myanmar	Insectivorous.
Megadermatidae	False vampire bats	6 genera	6 species	Old World tropics to northern Australia	Insectivorous, some partial carnivory.
Rhinonycteridae	Trident bats	4 genera	9 species	Gulf of Oman south through (mainly East) Africa to southern Africa, northwest Australia.	Insectivorous.
Hipposideridae	Old World leaf-nosed bats	7 genera	88 species	Old World tropics and subtropics	Insectivorous.
Rhinolophidae	Horseshoe bats	1 genus	109 species	Old World north to c.55° in Europe and 45° in Japan and south throughout Africa and the eastern edge of Australia	Insectivorous.
Emballonuridae	Sheath-tailed bats	14 genera	54 species	Through Old and New World tropics and subtopics north to Turkey and south to Victoria, Australia	Insectivorous.
Nycteridae	Slit-faced bats	1 genus	15 species	Africa, two species Southeast Asia	Insectivorous, some partly carnivorous.
Myzopodidae	Madagascar sucker-footed bats	1 genus	2 species	Madagascar	Insectivorous.
Mystacinidae	Short-tailed bats	1 genus	2 species	New Zealand	Insectivorous, with some fruits and flowers.
Noctilionidae	Fisherman or bulldog bats	1 genus	2 species	Central and South America	Insectivorous and carnivorous (fish).
Furipteridae	Smoky bat and thumbless bat	2 genera	2 species	Central America to southern Brazil, N Chile	Insectivorous.
Thyropteridae	Disc-winged bats	1 genus	5 species	Central and South America south to central/southern Brazil	Insectivorous.
Mormoopidae	Ghost-faced or moustached bats	2 genera	18 species	Central and South America, just into North America	Insectivorous.

Family	Common name	No. of genera	No. of species	Distribution	Food
Phyllostomidae	New World spear-nosed bats	60 genera	217 species	Southern North America south to c.35˚S in Argentina and Chile	Insectivorous, fruit- and flower-feeding, a few species carnivorous or sanguinivorous.
Natalidae	Funnel-eared bats	3 genera	12 species	Caribbean, Mexico to Brazil	Insectivorous.
Molossidae	Free-tailed bats	22 genera	126 species	To c.40–45˚N from southern France in the west to Korea and Japan in the east and similar latitudes in North America, south to southern Australia and throughout South America to c.45˚S	Insectivorous.
Miniopteridae	Bent-winged or long-fingered bats	1 genus	38 species	Old World tropics and subtropics, north to southern Europe and Japan, south to South Africa and south Australia	Insectivorous.
Cistugidae	Wing-gland bats	1 genus	2 species	Africa	Insectivorous.
Vespertilionidae	Plain-nosed or vesper bats	54 genera	496 species	Worldwide, north to the Arctic Circle, south to southernmost South America, Africa and Australia and New Zealand	Insectivorous.

Further information

Altringham, J.D. 1996. *Bats – Biology and Behaviour.* Oxford
University Press, Oxford, 262pp.

Crichton, E.G. & Krutzsch, P.H. 2000. *Reproductive Biology
of Bats.* Academic Press, London, 510pp.

Fenton, M.B. & Simmons, N.B. 2014. *Bats – A World of Science
and Mystery.* University of Chicago Press, Chicago, 240pp.

Gunnell, G.F. & Simmons, N.B. 2012. *Evolutionary History
of Bats: Fossils, Molecules and Morphology.* Cambridge
University Press, Cambridge, 560pp.

Kunz, T.H. & Fenton, M.B. (eds.) 2003. *Bat Ecology.* The
University of Chicago Press, Chicago, 779pp.

Kunz, T.H. & Parsons, S. (eds.) 2009. *Ecological and
Behavioral Methods for the Study of Bats.* The John Hopkins
University Press, Baltimore, 901pp.

Hutson, A.M., Mickleburgh, S.P. & Racey, P.A. (comp.) 2001.
*Microchiropteran Bats: global status survey and conservation
action plan.* IUCN/SSC Chiroptera Specialist Group.
IUCN, Gland, Switzerland and Cambridge, UK, 258pp.

Mickleburgh, S., Hutson, A.M. & Racey, P.A. (eds) 1992.
Old World Fruit Bats - an action plan for their conservation.
IUCN, Switzerland, 252pp.

Neuweiler, G. 2000. *The Biology of Bats.* Oxford University
Press, Oxford, 310pp.

Novak, R.M. 1994. *Walker's BATS of the World.* The John
Hopkins University Press, Baltimore, 287pp.

Russ, J. (ed.) 2021. *Bat Calls of Britain and Europe.* Pelagic
Press, Exeter, 462pp.

Scheeburger, K. & Voigt, C.C. 2016. *Zoonotic Viruses and
Conservation of Bats.* Chapter 10 (pp.263-292) in Voigt,
C.C. & Kingston, T. (eds.) 2016. *Bats in the Anthropocene:
Conservation of Bats in a Changing World.* Springer Open,
Heidelberg, 606pp.

Taylor, M. & Tuttle, M. 2018. *Bats – An illustrated guide to all
species.* Smithsonian Books, Washington, 400pp.

Teeling, E.C., Vernes, S.C., Davalos, L.M., Ray, D.A.,
Gilbert, M.T.P., Myers, E. & Bat1K Consortium 2018.
Bat Biology, Genomes, and the Bat1K Project: To
Generate Chromosome-level Genomes for All Living Bat
Species. *Annual Review of Animal Biosciences* 6: 23–46.

Voigt, C.C. & Kingston, T. (eds.) 2016. *Bats in the
Anthropocene: Conservation of Bats in a Changing World.*
Springer Open, Heidelberg, 606pp.

Wilson, D.E. & Mittermeier, R.A. (eds) 2019. *Handbook
of the Mammals of the World, Vol 9. Bats.* Lynx Edicions,
Barcelona, 1008pp.

ORGANISATIONS

Bat Conservation International
https://www.batcon.org/

Bat Conservation Trust, UK
https://www.bats.org.uk/

IUCN/SSC Bat Specialist Group
https://www.iucnbsg.org/

IUCN Red List
https://www.iucnredlist.org/

Africa:
Bat Conservation Africa
https://www.facebook.com/batconafrica/

Australasia:
Australasian Bat Society (ABS)
http://www.ausbats.org.au

Europe:

BatLife Europe
https://www.batlife-europe.info/

Eurobats – Agreement on the Conservation of Populations
of European Bats
https://www.eurobats.org/

Latin America::

Chiroptera Neotropical
http://chiroptera.conservacao.org/Home.html

Red LatinoAmericana y del Caribe para la Conservacion
de los Murcielagos (RELCOM)
http://www.relcomlatinoamerica.net/

North America (Canada, United States, Mexico):
[North American Bat Conservation Partnership (NABCP)
http://www.batcon.org/nabcp/newsite/index.html

North American Bat Conservation Alliance (NABCA)
https://batconservationalliance.org/

South Asia:
Chiroptera Conservation and Information Network of South
Asia (CCINSA)
http://www.zooreach.org/Networks/Chiroptera/Chiroptera.htm

Southeast Asia:
Southeast Asian Bat Conservation Research Unit (SEABCRU)
http://www.seabcru.org/

Index

Picture credits

Acknowledgement

My thanks to Professor Paul Racey who answered many of my questions in the preparation of this book and for his long-term support in my involvement with bats. Others unwittingly helped with discussion on particular matters, such as Peter Scrimshaw and David King on echolocation. My wife, Jacqui, has been very tolerant of the long periods of me locked away in my study. And I thank the publications team at the Natural History Museum, London, for their encouragement and guidance through the publishing procedure.